D0906304

Foundations of Anatomy and Physiology

Janet S. Ross RGN RFN RNT Sister Tutor Certificate
(University of Edinburgh)

Formerly Director of Nurse Training, North Edinburgh School of Nursing,
Western General Hospital, Edinburgh
Examiner to the General Nursing Council for Scotland
Formerly Departmental Sister, Orthopaedic Department,
Royal Infirmary, Glasgow
Nursing Sister and Sister Tutor, Colonial Nursing Service, Gold Coast

Kathleen J. W. Wilson BSc PhD RGN SCM RNT

Senior Research Fellow, University of Birmingham
Formerly Senior Lecturer, Department of Nursing Studies,
University of Edinburgh
Formerly Principal Tutor, Preliminary Training School, Royal Infirmary,
Edinburgh
Midwife, University College Hospital, London
Nursing Sister, Colonial Nursing Service, Gold Coast

 Livingstone Nursing Texts

Foundations of Anatomy and Physiology

FOURTH EDITION

Janet S. Ross
Kathleen J. W. Wilson

Pfeiffer Library
Pfeiffer College
Misenhelmer, N. C. 28109

DISCARD

113138

Churchill Livingstone
Edinburgh London and New York 1973

CHURCHILL LIVINGSTONE
Medical Division of Longman Group Limited

Distributed in the United States of America by Longman
Inc., 19 West 44th Street, New York, N.Y. 10036 and
by associated companies, branches and representatives
throughout the world.

© Longman Group Limited 1973

All rights reserved. No part of this publication may be
reproduced, stored in a retrieval system, or transmitted in
any form or by any means, electronic, mechanical, photo-
copying, recording or otherwise, without the prior permis-
sion of the publishers (Churchill Livingstone, 23 Ravelston
Terrace, Edinburgh EH4 3TL).

First Edition 1963
  Reprinted 1964, 1965
Second Edition 1966
  Reprinted 1967
Third Edition 1968
  Reprinted 1968, 1969, 1970, 1971
Fourth Edition 1973
  Reprinted 1975
  Reprinted 1976
  Reprinted 1977
  Reprinted 1978

ISBN 0 443 01087 0

Printed in Hong Kong by
Sheck Wah Tong Printing Press

Preface to Fourth Edition

In preparing the fourth edition of this book, considerable changes have been made in the text and in the arrangement of the material.

These are the result of our experience as teachers of this subject and of the changes in nursing practice which have occurred. Much of the text has been rewritten in order to clarify and make additions to physiological explanations and to remove unnecessary anatomical detail.

We extend our thanks to Mr W. G. Henderson, Miss Mary C. Emmerson and other members of the staff of Churchill Livingstone for their help and guidance throughout the preparation of this edition.

Edinburgh, 1973   J. S. Ross
Birmingham, 1973   K. J. W. Wilson

Preface to First Edition

In writing this book we have tried to introduce the subject of Anatomy and Physiology in as simple a form as possible. The subject matter has been arranged to cover Anatomy and Physiology of the Syllabus of the General Nursing Council for Scotland, England and Wales and Northern Ireland.

We have introduced, as far as possible, up-to-date anatomical terminology and have excluded outdated nomenclature. We have tried to illuminate the text with comprehensive diagrams and, for clarity, many of these are coloured. It is to be hoped that the student will be able to acquire a sound knowledge of the subject, by using these diagrams, by examining and handling anatomical models and bones, as well as from the text.

We hope that this book may not only be of value to Student Nurses but to Staff Nurses and Ward Sisters as well as to Nurses who are attending Ward Sister, Clinical Instructor and Nurse Teaching Courses.

We wish to express our grateful thanks to all those who have helped in the preparation of this book: Miss C. F. MacKinnon, Midwifery Tutor, for her suggestions on the Female Reproductive System; Mr Charles Macmillan of E. & S. Livingstone Ltd., for his encouragement and guidance; Mr James Parker for his patience and constant helpful advice; Mr R. W. Matthews and Mr J. Gordon for their excellent diagrams all of which are line drawings and which we feel enhance the text.

J. S. Ross
K. J. W. Wilson

Edinburgh, 1962

Contents

Contents

Introduction

As nurses concern themselves with the prevention of illness as well as the care of people who have become ill, it is necessary that they should have a knowledge of the structure of the human body and how it functions in health. A knowledge of the normal is essential before there can be understanding of the abnormal.

Anatomy is the name given to the parts which make up the body, their arrangement, relationship to each other, and the different types of cells recognisable on microscopic examination. *Physiology* on the other hand is the name given to the methods by which these structures function and interact.

For descriptive purposes the body has been divided into systems, but it must be understood that no one system can function in isolation. Each one is constantly being influenced by environmental conditions.

In the wider subject of general biology there are single cell organisms which are capable of living an independent life, for example, the amoeba. Although this text begins with a description of cells, these cells are not capable of independent existence, but are units of the total complex human organism. These units have become specialised and many cells which have similar properties are grouped into tissues. An organ, such as the stomach, is composed of several different types of tissue and when a number of organs have related functions they are classified as a system.

It is to be hoped that this text will stimulate the interest of students to continue their studies at a more advanced level.

1. The Cells and Tissues of the Body

The human body develops from a single initial cell which is formed by the fusion and fertilisation of the female germ cell (the ovum) by the male germ cell (the spermatozoon). This initial cell is known as a *zygote*. The zygote grows and reproduces, forming millions of cells which develop into tissues. These tissues in turn develop into organs which form a new being.

The Structure of a Human Cell

All living cells are made of a substance known as *protoplasm* which has been described as the 'material or physical basis of all forms of life'.

Protoplasm is described as a slightly opaque colourless soft jelly-like substance consisting of water and the following substances in solution or suspension:

    organic and inorganic salts
    glucose
    lipids (fatty substances)
    nitrogenous substances.

The protoplasm of the cell is surrounded by a *cell membrane* which is semi-permeable. It is composed of protein threads and lipids. Between the threads and the lipids there are minute spaces termed 'pores' through which minute molecules can pass into the protoplasm. Slightly larger molecules of nutrient material may be dissolved in the lipids of the cell membrane and then transferred to the protoplasm. Still other nutrients may have to be actively transported across the cell membrane by chemical substances known as 'carriers'. Thus three methods are involved in supplying nutrition to the cell:

    by diffusion through the 'pores'
    by dissolving in the lipids of the cell membrane
    by the action of 'carriers'.

Within the heart of the cell is a central globular mass known as the *nucleus*. The nucleus is surrounded by a membrane, *the nuclear membrane*. The protoplasm between the nuclear membrane and the cell membrane is known as *cytoplasm* and that within the nucleus as *nucleoplasm*.

The cytoplasm contains protein molecules known as *ribonucleic acids* (RNA) and small granular structures called *mitochondria*. Mitochondria are thought to be involved in oxidative reactions which take place in the cell and to act as store houses for nutrient materials required to replace worn out cytoplasm. Within the cytoplasm there are clear circular spaces called *vacuoles*. The vacuoles may contain waste materials or secretions which the cell cytoplasm has formed. Near to the nucleus is to be seen a small spherical body, the *centrosome* which is surrounded by a radiating thread-like structure. The centrosome contains two dark minute circular bodies the *centrioles* which participate in the early stages of cell division.

The nucleus is composed of nucleoplasm. Within the nucleus are spherical *nucleoli* and *chromatin* threads which carry the *genes*. The characteristic compounds of the nuclei are *deoxyribonucleic acids* (DNA) which are the genetically inherited information required for the maintenance of cells and their reproduction.

When the cell is reproducing the genes become organised into elongated masses which are arranged in pairs. These are now known as the *chromosomes*. Each cell of the human body has 46 chromosomes arranged in 23 pairs.

Figure 1 illustrates a circular cell but it must be understood that the cells forming the human body vary considerably in *shape and size* depending upon their function.

Figure 1

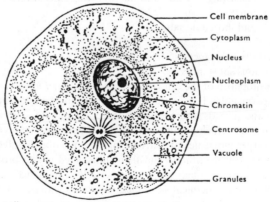

A simple cell.

Cell division

The cells of the human body reproduce or divide in a complicated fashion known as *mitosis*. Mitosis can be described as occurring in several stages or phases.

Figure 2

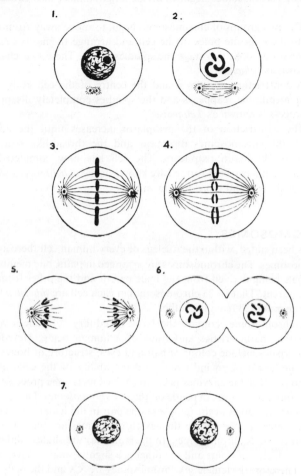

Diagram illustrating mitosis.

1. The centrosome divides into two, and each centrosome contains a centriole. The two centrosomes migrate to opposite poles of the cell but remain attached by fine thread-like spindles. This stage is known as *Prophase*. The chromatin becomes concentrated and more definite in shape and forms dark rod-shaped structures known as the *chromosomes*. There are 46 of these chromosomes, and they contain the genes which determine the inherited traits and characteristics of the individual.

2. The nuclear membrane now disappears and the chromosomes arrange themselves round the centre of the cell and appear to be attached to the spindles of the centrosome which are now at either pole of the cell. These two changes are known as *Metaphase*.

3. The chromosomes begin to divide longitudinally into two equal parts.

4. The two groups of chromosomes begin to move away from each other to the opposite poles of the cell and arrange themselves round the centrosomes. At this stage the spindles break. These two changes are known as *Anaphase*.

5. A constriction appears round the centre of the cell body. The nuclear membrane reappears and the spindles completely disappear. This process is known as *Telophase*.

6. The constriction of the cytoplasm increases until the cell is divided, the chromosomes disappear and the thread-like structure known as chromatin reappears. The cell has now completed its division and the two cells formed are known as the *daughter cells* which in turn will grow and reproduce by mitosis.

CHROMOSOMES

As has been stated, within the nucleus of every human cell there are 46 chromosomes. The chromosomes are arranged in pairs, one member of each pair being derived from the male parent and the other from the female parent. That is, 23 chromosomes in each cell are *inherited* from the mother and 23 from the father.

The chromosomes contain the basic hereditary substances which decide the characteristics and traits of a human being. Hereditary characteristics include colour of hair and eyes, structure of bones and teeth, the height of an individual and the ability of the cells of the body to produce the enzymes necessary for its metabolic processes.

Determination of sex depends on the *sex chromosomes*. One pair of the 23 chromosomes in the male and one pair in the female are the sex chromosomes. In the female the sex chromosomes are the same and are termed the X chromosomes. In the male they are slightly different, one is an X chromosome and the other is a slightly smaller Y chromosome. Therefore the female sex chromosomes are XX and the male XY.

In the ovum or female egg cell and in the spermatozoon or male germ cell the number of chromosomes is reduced to 23 single chromosomes. This means that all ova have an X chromosome and half the spermatozoa have an X chromosome and half a Y chromosome.

In conception, if an X bearing spermatozoon fertilises an ovum the offspring will be female and if a Y bearing spermatozoon fertilises an ovum the offspring will be male.

Sperm X + ovum = offspring XX = female.
Sperm Y + ovum = offspring XY = male.

When a pair of genes is being considered it is found that one may exert a stronger influence than the other. The gene exerting the stronger influence is termed *dominant* and the gene which is less effective is described as *recessive*. The characteristics of the offspring such as height, colour of eyes and hair and other familial characteristics depends upon the dominance of the genes.

Properties and Functions of Cells

Cells are living structures, therefore they exhibit the following properties which are common to living things.

1. Metabolism

This is a general term applied to the various changes, whatever their nature, taking place in living cells. For example, the taking in and utilisation of nourishment:

(a) to produce heat and energy
(b) to build up and repair protoplasm which becomes worn out during other functions
(c) to produce secretions or enzymes.

The cell receives its nourishment from the blood stream, and this nourishment passes into the cell through its semi-permeable membrane.

2. Respiration

Every cell requires oxygen for the process of metabolism. The oxygen is absorbed through the semi-permeable wall and is used for the oxidation of nutrient materials thus providing heat and energy. Waste products are produced, for example, carbon dioxide and water, which are passed out from the cell through its semi-permeable wall. The utilisation of oxygen and production of carbon dioxide is known as *cellular respiration.*

3. Growth

Cells have the ability to grow until they are mature and ready to reproduce.

4. Excretion

During metabolism various substances are produced which are of no further use to the cell and may cause damage if they are retained. These waste products are excreted through the semi-permeable cell membrane.

5. Movement

Movement of the whole or part of a cell may occur. White blood cells, for example, are able to move freely. The movement described in other cells may simply be a continual movement of the protoplasm, the movement of cilia, or the contraction of a muscle fibre.

6. Irritability

This means that the cell can respond to a stimulus which may be physical, chemical or thermal. For example, a muscle fibre contracts when stimulated by a nerve impulse.

7. Reproduction

As mentioned previously cells grow and when growth is completed reproduction takes place.

The Body as a Whole

The cell is the unit of structure of the body, and the body consists of millions of cells which can only be seen with the aid of a microscope. As development proceeds groups of cells become differentiated from one another and are built up into different patterns to form the *tissues* of the body. In turn the tissues arrange themselves into patterns which are known as *organs* and certain groups of organs form *systems*. The human body can be described in various ways but here it shall be described systematically. It must be remembered that these systems cannot be considered as independent entities as each one is necessary for the functioning of the others.

The Elementary or Fundamental Tissues of the Body

There are four elementary or fundamental tissues which make up the body as a whole:

    epithelial tissue or epithelium
    connective tissue
    muscle tissue
    nervous tissue.

THE EPITHELIAL TISSUES

The cells forming these tissues are very closely packed together and the inter-cellular substance, known as the *matrix*, is reduced to a minimum. The cells usually lie on a *basement membrane* from which they receive their nourishment.

Epithelial tissue is divided into two types: *simple* and *compound*.

SIMPLE EPITHELIUM

Simple epithelium consists of a single layer of cells and is divided into several types. The types are named according to the shape of the cells which differ according to the functions they perform.

Squamous or pavement epithelium

Squamous epithelium is composed of a single layer of flattened cells. The cells fit closely together like flat stones forming a pavement. In this way a very smooth surface is formed.

The function of squamous epithelium is to provide a thin smooth inactive lining for the following organs:

    the heart
    the blood vessels
    the alveoli of the lungs
    the lymphatic vessels.

Figure 3

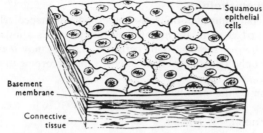

Squamous epithelium.

Cubical or cuboidal epithelium

Cubical epithelium is composed of cube-shaped cells fitting closely together. These cells are found forming the tubules of the kidneys and some of the glands, for example, the thyroid gland. Its function is secretory.

Figure 4

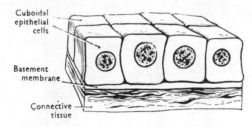

Cuboidal epithelium.

Columnar epithelium

This is formed by a single layer of cylindrical-shaped cells. These cells also produce a secretion and form the lining to the following organs:

the stomach
the small intestine
the large intestine
the gall bladder and bile ducts.

Figure 5

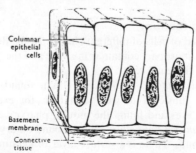

Columnar epithelium.

Ciliated epithelium

Ciliated epithelium is formed also by cylindrical-shaped cells but on their free edges there are minute hair-like processes known as *cilia*. The function of the cilia is to perform a lashing movement in one direction only. The effect of their activity varies according to the site of the tissue.

1. Forming the inner lining of:

the nose

the larynx

the trachea

the bronchi.

The function of the cilia is to waft mucus and dust or other small particles from these organs towards the throat. The movement of the cilia is likened to the movement in a field of corn which has been stirred by a breeze.

2. Lining the uterine tubes where its function may be to assist in propelling the ovum towards the uterus.

Figure 6

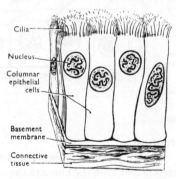

Ciliated epithelium.

COMPOUND EPITHELIUM

Compound epithelium consists of many layers of cells and is divided into several types.

Stratified epithelium

This is composed of cells of different shapes. In the deepest layers the cells are for the most part columnar in shape and the superficial layers are made up of flattened cells.

Non-keratinised Stratified Epithelium. This is found on wet surfaces which may be subjected to wear and tear, for example, the conjunctiva of the eyes and lining the mouth, pharynx and oesophagus.

Keratinised Stratified Epithelium. This is found on dry surfaces, that is skin, hair and nails. There is a surface layer of keratin which prevents injury to and drying of the underlying cells.

Figure 7

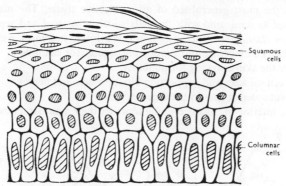

Squamous
cells

Columnar
cells

Stratified epithelium.

Transitional epithelium

This is composed of several layers of pear-shaped cells. The superficial layer of cells may be flattened. This tissue is found:

in the pelvis of the kidney
lining the ureters
lining the urinary bladder.

Figure 8

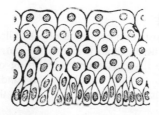

Transitional epithelium.

CONNECTIVE TISSUES

The cells forming the connective tissues are more widely separated than those forming the epithelial tissues and the inter-cellular substance is consequently increased in amount. There may or may not be fibres present in the matrix. The matrix may be of a semi-solid jelly-like consistency or dense and rigid depending upon the position and function of the tissue.

The connective tissues are sometimes described as the supporting tissues of the body because their functions are mainly mechanical, connecting together more active tissues into functional units.

AREOLAR TISSUE

This is the most generalised of all connective tissue. The matrix is described as semi-solid with the cells widely dispersed and separated by *yellow elastic fibres* and *white fibres*. It is found in almost every part of the body connecting and supporting organs, for example:

under the skin allowing for pliability

between muscles and supporting blood vessels and nerves

as a submucous coat in the digestive tract

in the interior of organs binding together their main structure.

Figure 9

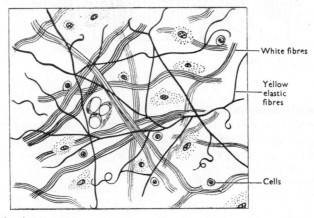

White fibres

Yellow elastic fibres

Cells

Areolar tissue.

ADIPOSE TISSUE

Adipose tissue consists of a collection of cells filled with fat globules known as *fat cells*. These cells are present in the matrix of areolar tissue.

Fatty or adipose tissue is found supporting organs such as the kidneys and the eyes. It is also found between bundles of muscle fibres, and with areolar tissue under the skin giving the body a smooth continuous outline.

Figure 10

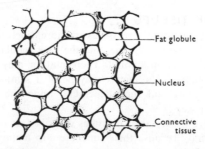

Fat globule

Nucleus

Connective tissue

Fatty or adipose tissue.

WHITE FIBROUS TISSUE

This is a strong connecting tissue made up mainly of closely-packed bundles of *white fibres* with very little matrix. There are very few cells and these lie in rows between the bundles of fibres. Fibrous tissue is found:

forming the ligaments binding bones together

as an outer protective covering for bone, known as the *periosteum*

as an outer protective covering of some organs, for example, the kidneys, lymphatic glands, blood vessels, the brain, and

forming muscle sheaths called *muscle fascia.*

Figure 11

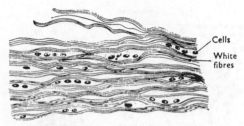

Cells

White fibres

White fibrous tissue.

YELLOW ELASTIC TISSUE

Yellow elastic tissue is capable of considerable extension, and recoil. There are few cells and the matrix consists mainly of masses of *yellow elastic fibres.* It is found in organs where alteration of shape is required, for example:

in the arteries, particularly the large arteries

in the trachea and bronchi

in the lungs.

Figure 12

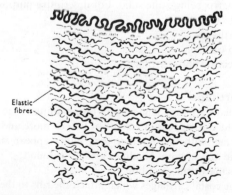

Elastic fibres

Yellow elastic tissue.

LYMPHOID TISSUE

This tissue has a semi-solid matrix with fine branching fibres, the cells are specialised cells known as *lymphocytes*. Lymphoid tissue is found:

in the lymphatic nodes
in the spleen
in the tonsils and the adenoids
in the vermiform appendix
forming the solitary glands in the small intestine.

Figure 13

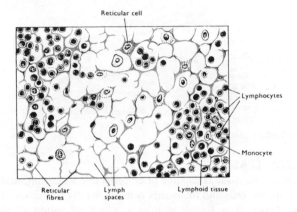

Lymphoid tissue.

CARTILAGE

Cartilage is a much firmer tissue than any of the other connective tissues; the matrix being quite solid. For descriptive purposes cartilage is divided into three types:

hyaline cartilage
white fibro-cartilage
yellow or elastic fibro-cartilage.

Hyaline cartilage

Hyaline cartilage appears as a smooth bluish-white tissue. Under the microscope the cells appear in groups of two or more and where they come in contact with one another their edges appear straight. The matrix is solid and smooth. Hyaline cartilage is found:

on the articular surfaces of bones
forming the costal cartilages which attach the ribs to the sternum
forming part of the larynx, trachea and bronchi.

Figure 14

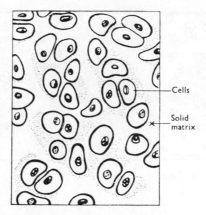

Cells

Solid
matrix

Hyaline cartilage.

White fibro-cartilage
This consists of dense masses of white fibres in a solid matrix with the cells widely dispersed. It is a dense, tough, slightly flexible tissue. White fibro-cartilage is found:
>   as pads between the bodies of the vertebrae called the inter-vertebral discs
>   between the articulating surfaces of the bones of the knee joint known as the semi-lunar cartilages
>   surrounding the rim of the bony sockets of the hip and shoulder joints deepening the cavities.

Figure 15

White
fibres

White fibro-cartilage.

Yellow or elastic cartilage
This consists of yellow elastic fibres running through the solid matrix. The cells lie between the fibres. It forms the pinna or lobe of the ear and the epiglottis.

Figure 16

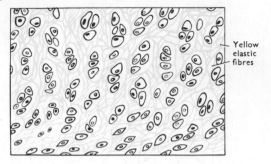

Yellow elastic cartilage.

BLOOD
This is a fluid connective tissue and will be described in detail later, in chapter 5.

BONE
Bone is one of the hardest connective tissues in the body and when developed is composed of:
    water, 25 per cent
    organic material, 30 per cent
    inorganic salts, 45 per cent.

There are two types of bone tissue described:
compact bone tissue and *cancellous bone tissue*.

Compact bone
To the naked eye compact bone appears to be a solid structure, but when examined under a microscope definite patterns can be seen. These patterns are termed the *haversian systems* (Fig. 17).

A haversian system has several well-defined characteristics.

1. A central *haversian canal* which runs longitudinally. Traversing the haversian canal are minute arterial capillaries, venous capillaries, lymphatic capillaries and nerves.
2. Surrounding this central canal are concentric rings of flat bone known as the *lamellae*.
3. Between the lamellae are spaces or *lacunae* containing lymph and bone cells.
4. Running between the lacunae and the central haversian canal are fine channels called the *canaliculi*. Lymph carrying nourishment to the bone cells flows through these canals.
5. In the spaces between the haversian system there are minute circular plates of bone known as the *interstitial lamellae*.

Figure 17A

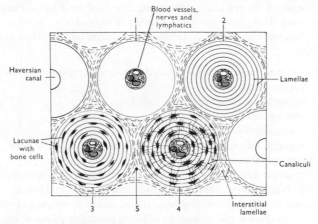

Diagrammatic illustration of the microscopic structure
of bone. A cross section.

Cancellous bone

Cancellous bone, to the naked eye, appears like a sponge. Microscopically the haversian canals are much larger and there are fewer
lamellae, giving the appearance of a honeycomb. *Red bone marrow* is
always present within cancellous bone.

Figure 17B

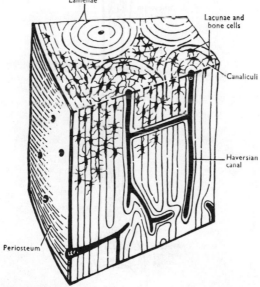

A longitudinal section.

Red Bone Marrow. This is known as the blood forming tissue because in this marrow the red blood cells, white blood cells and blood platelets are formed. These cells grow and develop in the red bone marrow and only when mature do they enter the blood stream and circulate round the body.

The outer surface of bone is covered by a vascular fibrous membrane known as the *periosteum*.

Peristeum has several very important functions:

it forms an outer protective covering

it gives attachment to the muscle tendons

it is through the periosteum that blood vessels pass to the bone to provide its blood supply

in its deeper layer there are many bone forming cells (*osteoblasts*) which are responsible for the deposition of new bone tissue.

Periosteum is found covering bone tissue, except those parts of bones which participate in the formation of freely moveable joints where the periosteum is replaced by *articular* or *hyaline cartilage*.

Figure 18

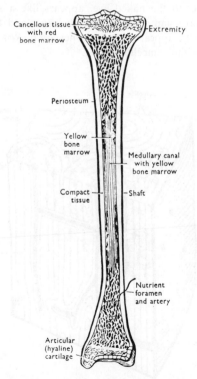

Cancellous tissue with red bone marrow

Extremity

Periosteum

Yellow bone marrow

Medullary canal with yellow bone marrow

Compact tissue

Shaft

Nutrient foramen and artery

Articular (hyaline) cartilage

Diagram of the naked-eye structure of a long bone, longitudinal section.

The *skeleton* or bony framework of the body is made up of a variety of bones which are classified as follows:

long bones
irregular bones
short bones
flat bones
sesamoid bones.

These bones are composed of both *compact* and *cancellous* bone tissue. The relative quantity of these two types of tissue in each bone varies according to the need for strength or lightness.

Long bones

When long bones are examined by the naked eye they are described as having a *shaft* or diaphysis and *two extremities* or epiphyses.

The Shaft. This is composed mainly of compact bone with a central canal known as the *medullary canal.* In adult life this canal is filled with adipose tissue which is known as the *yellow bone marrow.*

The Extremities. These are composed of a thin layer of compact bone covered by hyaline cartilage within which there is cancellous bone tissue containing red bone marrow.

Irregular, short, flat and sesamoid bones

These are composed of a thin layer of compact bone surrounding an inner mass of cancellous bone containing red bone marrow and, like long bones, they are encased in periosteum except on their articular surfaces.

Figure 19

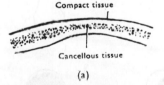

Compact tissue

Cancellous tissue

(a)

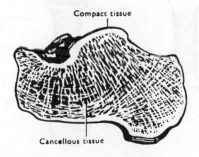

Compact tissue

Cancellous tissue

(b)

Diagram of the structure of a flat and irregular bone.

The development of bones

The bones forming the skeleton begin developing before birth and this development is not complete until about the twenty-fifth year of life. Long bones develop from *rods of cartilage* and the other varieties of bones develop from *membrane*, with the exception of sesamoid bones which develop from *tendon*.

There are two types of bone cells involved in the development and repair of bone: *osteoblasts* and *osteoclasts*.

The osteoblasts are responsible for *building* bone.

The osteoclasts have an opposing action and are responsible for *destroying* bone. Due to the function of the osteoclasts the *medullary canals* of long bones are formed and the *sinuses* (hollow cavities) within many bones are created. The osteoclasts are also responsible for shaping and remodelling bones during their development and subsequent repair.

Development of Long Bones. Long bones develop by a complicated series of changes causing the destruction of cartilage and its replacement by bone tissue. About the eighth week of fetal life a *primary centre of ossification* appears in the middle of the rod of cartilage and,

Figure 20

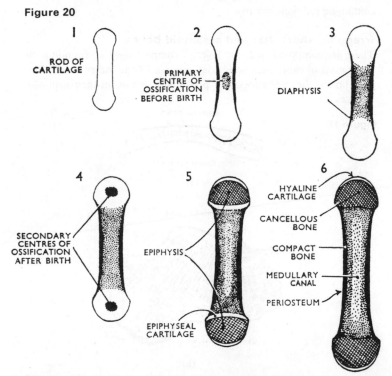

Diagrammatic illustration of the development of a long bone.

by the action of enzymes on the bone cells (osteoblasts) of the primary centre, bone formation takes place together with the deposition of mineral salts, mainly calcium phosphate.

From this primary centre of ossification the *diaphysis* or *shaft* of the bone is formed and ossification spreads upwards and downwards from this point. The development of the shaft in circumference occurs due to the laying down of bone cells and inorganic salts by the osteoblasts which are present in the periosteum.

Generally after birth *secondary centres of ossification* appear in the part of the cartilage which will eventually be the *epiphyses* or *extremities* of the bone. From these centres ossification spreads upwards, outwards and downwards.

Bone grows in length at the *epiphyseal cartilage* which is found between each epiphysis and the diaphysis. The growth of the bone is not complete until the rate of ossification overtakes the rate of growth of the epiphyseal cartilage. Until ossification is complete a thin clear 'epiphysial line' shows on X-ray plates.

Flat, Short, Irregular and Sesamoid. These *bones* develop from one or more primary centres of ossification.

Function of bones

1. The bones of the skeleton form a supporting framework for the body.

2. Some of the bones give support to the weight of the body, for example, the bones of the lower limbs.

3. They form the boundaries for many of the body cavities, for example, the cranial cavity and the thoracic cavity.

4. They form a protection for the more delicate organs.

5. They provide the levers which are essential for the movement of the body.

6. They provide attachment for the voluntary muscles through the periosteum.

7. Due to the presence of red bone marrow in cancellous bone they have blood forming functions.

8. Due to the presence of calcium salts bone forms one of the stores for this substance.

MUSCLE TISSUE

There are three main types of muscle tissue described:

striated or *voluntary muscle*
smooth or *involuntary muscle*
cardiac muscle.

Striated or voluntary muscle

This may be described as *skeletal, striated, striped* or *voluntary* muscle. It is known as voluntary because it is under the control of the will.

When voluntary muscle is examined microscopically the cells are found to be roughly cylindrical in shape and of varying length (from 10 to 40 millimetres).

Each fibre has several nuclei situated just beneath the *sarcolemma* which is a fine sheath surrounding each muscle fibre. The fibres run parallel to one another and show well marked transverse dark and light bands, hence the name striated or striped muscle.

Figure 21

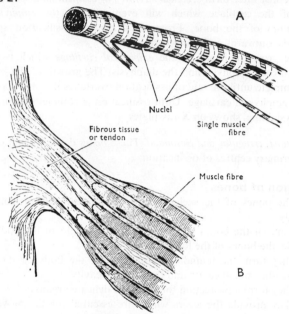

(A) Diagram of a group of muscle fibres.
(B) Diagram of a muscle and its tendon.

A muscle consists of a large number of muscle fibres. In addition to the sarcolemma mentioned previously, each fibre is enclosed in and attached to fibrous tissue called *endomycium*. Small bundles of fibres are enclosed in *perimycium* and the whole muscle in *epimycium*. The fibrous tissue enclosing the fibres, the bundles and the whole muscle extends beyond the muscle fibres to become the *tendon* which attaches the muscle to bone or skin.

Smooth or involuntary muscle
Smooth muscle may also be described as *involuntary, plain* or *visceral* muscle. It is not under the control of the will. It is found in the walls of blood and lymph vessels, in the alimentary canal in the respiratory tract, the urinary bladder and in the uterus.

Figure 22

Smooth or involuntary muscle cell.

When examined under a microscope the cells are *spindle* shaped with only one central nucleus. There is no distinct sarcolemma but a very fine membrane surrounds each fibre. The fibres are arranged together to form *bundles* and in turn the bundles are surrounded by areolar tissue to form *sheets* of muscle.

Cardiac muscle

This type of muscle tissue is found exclusively in the wall of the heart. It is not under the control of the will but varies considerably in structure from involuntary muscle.

When examined under a microscope cardiac muscle fibres are roughly cylindrical in shape and have a striated appearance. The fibres frequently divide and join other fibres or there are short communicating fibres. This division and interlacing of fibres is described as a *pseudo-syncytium mass*. There are clearly defined nuclei to be seen. Many fibres are arranged into bundles and these bundles are surrounded by connective tissues.

Figure 23

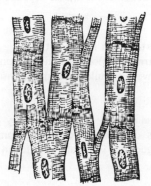

Cardiac muscle.

PROPERTIES OR CHARACTERISTICS OF MUSCLE TISSUE
Composition
20 per cent protein
75 per cent water
5 per cent mineral salts, glycogen, glucose and fat.

Functions

The function of muscle is to *contract* and *pull*. A muscle fibre will contract when it is stimulated by one of the following means:

electrical
mechanical
chemical
thermal.

In the human body the necessary stimulus is supplied by *nerves* which pass to the muscle and break up into minute nerve fibres each one of which stimulates a single muscle fibre. When the muscle fibre contracts it follows the *all or none law*, that means, it does not contract at all or it contracts to its full capacity and becomes *shorter* and *thicker*.

In order to contract when it is stimulated a muscle fibre must have a good *blood supply*. By this method the muscle fibres receive oxygen and nutritional materials and waste products are removed.

Muscle tone.

This is a state of partial contraction of muscles. It is achieved by the contraction of a few muscle fibres at a time. Muscle tone in relation to striated muscle is associated with the maintenance of posture in the sitting and standing positions. The muscle is stimulated to contract through a system of spinal reflex actions. Stretching the muscle or its tendon stimulates the reflex action (*see* Nervous System). A degree of muscle tone is also maintained by smooth and cardiac muscle.

Force of muscle contraction

The force of the contraction of the muscle depends upon the *number of fibres* stimulated.

Effect of temperature on muscle tissue

If muscle tissue is warm it responds to stimulation more rapidly and there is less delay between stimulation and response.

Muscle fatigue

If a muscle is stimulated to contract at very frequent intervals gradually it becomes depressed and will no longer contract. This is called muscle fatigue and is usually due to an inadequate blood supply.

Elementary chemistry of muscle contraction

The energy which muscles require is derived from the breakdown of carbohydrate and fat molecules inside the fibres. Each molecule undergoes a series of changes and, with each change, small quantities of energy are released. For the complete breakdown of these molecules and the release of all the available energy an adequate supply of oxygen is required. If the individual undertakes excessive exercise the oxygen supply may be insufficient, resulting in the accummulation of intermediate metabolic products such as lactic acid. Where the breakdown process is complete the waste products are carbon dioxide and water.

Further features of striated muscle

The voluntary muscles are the muscles which produce the movements of the body. Each muscle consists of a *fleshy part* made up of striped fibres and a *tendinous part,* usually at both ends of the fleshy part. It is by these tendons that the muscle is attached, usually, but not always, to bone. When the tendinous attachment of a muscle is broad and flat it is called an *aponeurosis.*

To be able to produce movement at a joint a muscle or its tendon *must stretch across the joint.* This is necessitated by the fact that when a muscle contracts its fibres *shorten and it pulls* which has the effect of pulling one bone towards another, for example, bending the knee.

The muscles of the skeleton are arranged in groups, some of which are *antagonistic* to each other. It is the state of muscle contraction within antagonistic groups of muscles which enables us to maintain an upright position; whereas to produce movement at a joint one muscle has to contract and its antagonist must relax. To carry out the fine movements of the fingers and arm in writing, there must be a very exact co-ordination of many muscles and groups of muscles.

As individual muscles and groups of muscles have names, it may be helpful to consider the origins of some of the names.

1. *The shape* of the muscle, for example, the deltoid.
2. *The direction* in which the fibres run, for example, the oblique muscles of the abdominal wall.
3. *The position* of the muscle, for example, the tibialis associated with the tibia.
4. *The movement* produced by contraction of the muscle, for example, flexors, extensors, adductors.
5. *The number of points of attachment* of a muscle, for example, the biceps muscle is so called because it has two tendons of origin.
6. *The names of the bones to which the muscle is attached,* for example, the carpi radialis and carpi ulnaris muscles.

NERVOUS TISSUE

The structure of nervous tissue is described in chapter 13 on the nervous system.

Membranes

Some of the tissues which line or cover organs are described as *membranes.*

The more important membranes can be classified as follows:
mucous membrane
serous membrane
synovial membrane.

Mucous membrane

This is the name given to the lining of the digestive tract, the respiratory tract and the genito-urinary tract.

The cells forming the membrane produce a secretion known as *mucus*. This is a slimy tenacious fluid containing a protein material known as *mucin*. This secretion is formed within the cytoplasm of the cells and as it accumulates the cells become distended and finally burst discharging the mucus on to the free surface. Thus the organs lined by mucous membrane have a moist slippery surface. As the cells fill up with mucus they have the appearance of a goblet or flask and are known as the *goblet cells*.

Figure 24

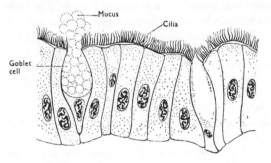

Diagram illustrating a goblet cell.

Serous membrane

Serous membranes consist of a double layer of tissue. The *visceral* layer closely invests the organ and the *parietal* layer lines the cavity in which the organ lies. The visceral layer is reflected off the organ to become the parietal layer and the two layers are separated by a watery or *serous fluid* secreted by the membrane.

There are three sites in which serous membranes are found. In the thoracic cavity the *pleura* surrounds the lungs and the *pericardium* the heart. In the abdominal cavity the *peritoneum* surrounds the abdominal organs.

The presence of serous fluid between the visceral and parietal layers of serous membrane enables an organ to move without being damaged by friction between it and adjacent organs. For example, the heart changes it shape and size during each beat and friction damage is prevented by the arrangement of pericardium and its serous fluid.

Synovial membrane

This membrane is found lining the joint cavities, surrounding tendons and ligaments. It is made up of a layer of fine flattened cells on a layer of delicate connective tissue.

Synovial membrane secretes a clear, sticky, oily fluid known as *synovial fluid*. This fluid acts as a lubricant to the joint and helps to maintain its stability. Synovial fluid also helps to nourish the articular cartilage and synovial cells remove damaging material from within the joint cavity.

Glands

Secreting epithelial cells line many of the organs of the body. The areas are usually vastly increased by invagination of the lining. When invagination occurs the structures formed are usually known as *glands*.

The glands vary in shape and complexity. The more complex glands require *canals* or *ducts* to carry the secretion to the surface. The names of the glands vary depending upon their complexity.

Simple tubular glands

The cells forming these glands form a single tube which opens directly on to the free surface.

Figure 25

A simple tubular gland.

Branched tubular glands

Here the deep part of the tube becomes branched having a more complex appearance.

Figure 26

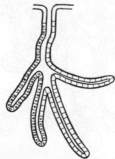

A branched tubular gland.

Simple saccular or alveolar glands

The cells forming these glands are arranged in a somewhat spherical formation surrounding a cavity which is known as the *saccule* or *alveolus*.

Figure 27

An alveolar gland.

Compound alveolar or racemose glands

The cells forming these very complex glands form many alveoli. Ducts carry away the secretion from each alveolus joining up with other ducts and finally empty the contents into one large duct which leads to the surface. The salivary glands are classical examples of *compound racemose glands*.

The glands described above are glands whose secretions are carried away by ducts. There are other glands found in the human body the secretions of which pass directly into the blood stream. These glands are known as the *ductless* or *endocrine glands* and will be described in a later chapter.

Figure 28

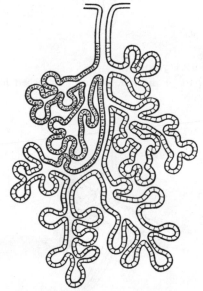

A compound racemose gland.

The Systems and Cavities of the Body

A *system* can be described as a group of structures or organs which together carry out essential and related functions.

The systems can be classified as follows:

osseus or skeletal	digestive
muscular	urinary
circulatory	nervous
respiratory	endocrine
	reproductive.

The body as a whole is built round the bony framework or skeleton and consists of a number of different parts:

the *head* and *neck*

the *trunk*, which can be divided into the *chest* or *thorax*, *abdomen* and *pelvis*;

the *limbs*, both upper and lower.

For descriptive purposes the human body is divided into *cavities* and the main organs of the body are contained in these cavities. There are four main cavities:

cranial	*abdominal*
thoracic	*pelvic.*

THE CRANIAL CAVITY

Figure 29

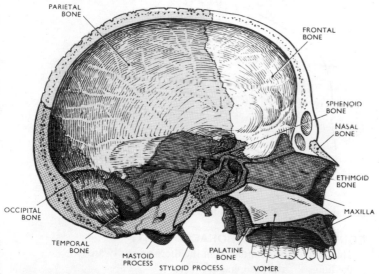

Diagram illustrating the boundaries of the cranial cavity.

The cranial cavity contains the *brain*, and its *boundaries* are formed by the bones of the cranium:

anteriorly	.	the frontal bone
laterally	.	the temporal bones
posteriorly	.	the occipital bone
superiorly	.	the parietal bones
inferiorly	.	the sphenoid, the ethmoid, and parts of the frontal, temporal and occipital bones.

THE THORACIC CAVITY

This cavity is situated in the upper part of the trunk. Its *boundaries* are formed by a bony framework and supporting muscles.

anteriorly	.	the sternum and costal cartilages of the ribs
laterally	.	twelve pairs of ribs and the intercostal muscles
posteriorly	.	the thoracic vertebrae and the intervertebral discs which lie between the bodies of the vertebrae
superiorly	.	the organs described as forming the root of the neck
inferiorly	.	the diaphragm, a dome-shaped muscle.

Figure 30

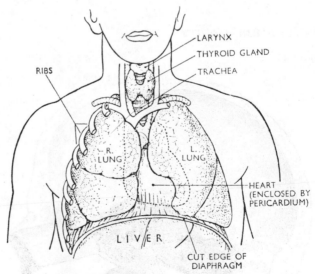

Diagram showing main organs in the thoracic cavity.

CONTENTS OF THE THORACIC CAVITY

The main organs contained in the thoracic cavity are:

the lungs	the oesophagus
the heart	the aorta (the largest artery in the body)

the trachea   the venae cavae (two large veins)
the bronchi   other blood vessels
               nerves, lymphatic vessels and glands.
The space between the lungs is known as the *mediastinum*.

THE ABDOMINAL CAVITY

This is the largest cavity in the body and is oval in shape. It is situated in the main part of the trunk and its *boundaries* are:

superiorly   .   the diaphragm

anteriorly   .   the muscles forming the anterior abdominal wall

posteriorly   .   the lumbar vertebrae and muscles forming the posterior abdominal wall

laterally   .   the lower ribs and parts of the mucles of the abdominal wall

inferiorly   .   the abdominal cavity is continuous with the pelvic cavity below.

Figure 31

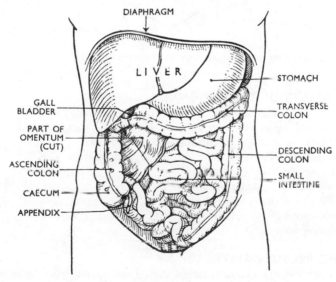

Diagram showing some of the organs in the abdominal cavity.

The abdominal cavity is divided into nine regions for descriptive purposes and for the purpose of locating the abdominal organs. This is done by the use of imaginary planes, two horizontal and two vertical. The zones thus formed are named in Figure 32.

Figure 32

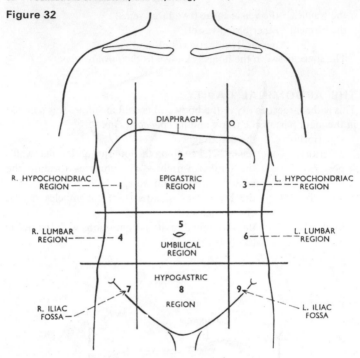

Diagram illustrating anatomical regions of the abdominal cavity.

CONTENTS OF THE ABDOMINAL CAVITY

Most of the organs which make up the digestive system lie in the abdominal cavity; the stomach, small intestine, part of the large intestine, the liver, the gall bladder and the pancreas.

Other organs present in the abdominal cavity are:
the spleen
the kidneys and the upper part of the ureters
the supra-renal glands
numerous blood vessels, nerves and lymphatic vessels
lymph nodes.

THE PELVIC CAVITY

The pelvic cavity is roughly funnel-shaped and extends from the lower end of the abdominal cavity. The *boundaries* of the pelvic cavity are:

superiorly	.	it is continuous with the abdominal cavity
anteriorly	.	the pubic bones and the symphysis pubis
posteriorly	.	the sacrum and coccyx
laterally	.	the bones of the pelvis
inferiorly	.	the levatores ani and coccygeal muscles forming the *pelvic floor*.

Figure 33

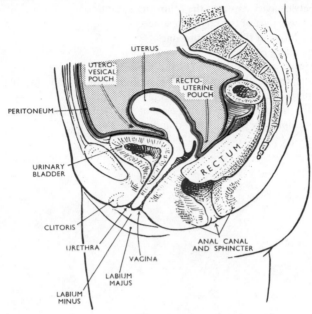

Diagram showing main organs in the female pelvic cavity.

Figure 34

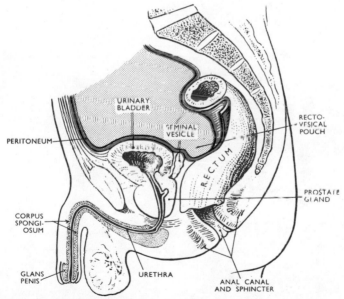

Diagram showing main organs in the male pelvic cavity.

CONTENTS OF THE PELVIC CAVITY

The pelvic cavity contains the following structures:

    the lower part of the large intestine

    some loops of the small intestine

    the urinary bladder

    the lower parts of the ureters and the urethra

    in the female the organs of the reproductive system; the uterus, uterine tubes, ovaries and vagina

    in the male some of the organs of the reproductive system, for example, the prostate gland and seminal vesicles.

2. The Skeleton

It has already been stated that there are two different types of bone tissue, cancellous and compact, and that they form bones which are classified as long bones, short bones, flat bones, irregular bones and sesamoid bones. All these types of bones form what is known as the *skeleton* or *bony framework* of the body. Before going on to discuss the individual bones which make up this framework it is necessary to be familiar with certain anatomical terms and their meaning.

The Anatomical Position. This is the position assumed in all anatomical descriptions The body is in the upright position with the head facing forward, the arms at the sides with the palms of the hands facing forward and the feet together.

Median Plane. When the body, in the anatomical position, is divided *longitudinally* into two equal parts it has been divided in the median plane. Any structure which is described as being *medial* to another is, therefore, nearer the midline and any structure which is *lateral* to another is farther from the midline or, at the side of the body.

Proximal and Distal. These terms are used when describing the bones of the limbs. The proximal end of a bone is that which is nearest the point of attachment of the limb, and the distal end that which is farthest away from the point of attachment of the limb.

Anterior or Ventral. This indicates that the part being described is nearer the front of the body.

Posterior or Dorsal. This means nearer the back of the body.

Superior. This indicates a structure nearer the head.

Inferior. This indicates a structure farther away from the head.

Border. This is a ridge of bone which separates two surfaces.

Spine, Spinous Process or Crest. This is a sharp ridge of bone.

Trochanter, Tuberosity or Tubercle. These are roughened bony projections usually for the attachment of muscles or ligaments. The different names are used according to the size of the projection. Trochanters are the largest and tubercles the smallest.

Styloid Process. This is a sharp downward projection of bone which gives attachment to muscles and ligaments.

Fossa (plural fossae). This is a hollow or depression in a bone.

Figure 35A

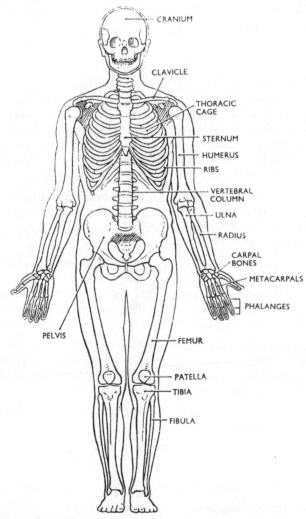

The skeleton—Anterior view.

Figure 35B

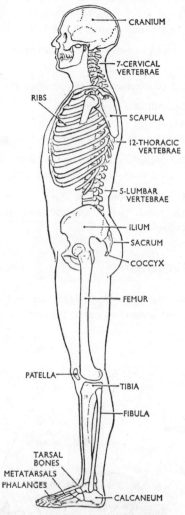

CRANIUM

7-CERVICAL
VERTEBRAE

RIBS

SCAPULA

12-THORACIC
VERTEBRAE

5-LUMBAR
VERTEBRAE

ILIUM

SACRUM

COCCYX

FEMUR

PATELLA

TIBIA

FIBULA

TARSAL
BONES
METATARSALS
PHALANGES

CALCANEUM

The skeleton—Lateral view.

Foramen (plural foramina). This is a hole in a bone.

Sinus. This is a hollow cavity within a bone.

Meatus. This indicates a tube-shaped cavity within a bone.

Articulation. This is a joint between two or more bones.

Suture. This is the name given to an immovable joint, e.g., between the bones of the skull.

Articulating Surface. This is the part of the bone which enters into the formation of a joint.

Facet. This is a small, generally rather flat, articulating surface.

Condyle. This is a smooth rounded projection of bone which takes part in a joint.

Septum. This is a partition separating two cavities.

Fissure or Cleft. This indicates a narrow slit.

The bones of the skeleton are divided into two main groups.

The axial skeleton consisting of the bones of the upright parts or axis of the body, i.e., the skull, vertebral column, ribs and sternum.

The appendicular skeleton consists of the shoulder girdles and the upper limbs and the pelvic girdles and the lower limbs. These are, therefore, the appendages.

The Axial Skeleton
THE SKULL
The skull rests upon the upper end of the vertebral column, and its bony structure is divided into two parts: the cranium and the face.

Figure 36

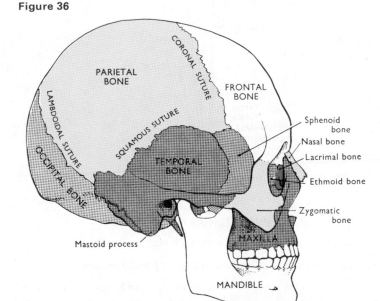

The bones of the skull—Lateral view.

THE CRANIUM
The cranium consists of eight bones.

Frontal bone	1
Parietal bones	2
Temporal bones	2
Occipital bone	1
Sphenoid bone	1
Ethmoid bone	1

The frontal bone
This is the bone of the forehead. It forms part of the orbital cavities and the prominent ridges above the eyes are called the *supra-orbital ridges*. These roughened ridges give attachment to the muscles which cause raising of the eyebrows. Just above the supra-orbital ridges, within the bone, there are two hollow spaces, or sinuses, which contain air, are lined with ciliated mucous membrane and communicate with the nasal cavities. The joint formed between the frontal bone and the parietal bones is called the *coronal suture*. The frontal bone forms other immovable joints with the sphenoid, zygomatic, lachrymal, nasal and ethmoid bones. The inner surface is grooved by the brain and blood vessels with which it is in close contact. At birth the bone consists of two parts which are separated by the *frontal suture*, but union is usually complete by the eighth year of life.

Parietal bones
These two bones form the sides and top of the skull. They articulate with each other at the *sagittal suture*, with the frontal bone at the coronal suture, with the occipital bone at the *lambdoidal suture* and with the temporal bones at the *squamous sutures*. The inner surface is concave and is grooved by the brain and blood vessels.

The temporal bones
These bones lie one on each side of the head and form immovable joints with the parietal, occipital and sphenoid bones. Each temporal bone is divided into four parts.

The Squamous Part. This is the fan-shaped portion which articulates with the parietal bone at the squamous suture.

The Mastoid Process. This is a thickened part of bone and can be felt just behind the ear. It contains a large number of very small air sinuses which communicate with the middle ear and are lined with simple squamous cells. A styloid process projects downwards from the mastoid process and gives attachment to muscles.

The Petrous Portion. This forms part of the base or floor of the skull and contains the organ of hearing.

Figure 37

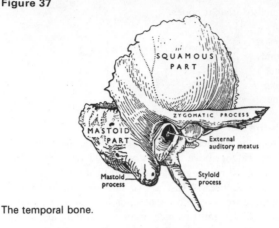

The temporal bone.

The Zygomatic Process. This is directed forward and articulates with the zygomatic bone to form the zygomatic arch.

The temporal bone has an articulating surface for the only movable bone of the skull, the mandible, at *the temporo-mandibular joint.* Immediately behind this articulating surface is the *auditory meatus* which passes inwards from the exterior towards the petrous portion of the bone.

The inner surface of the bone is deeply ridged by the brain and large blood vessels.

Figure 38

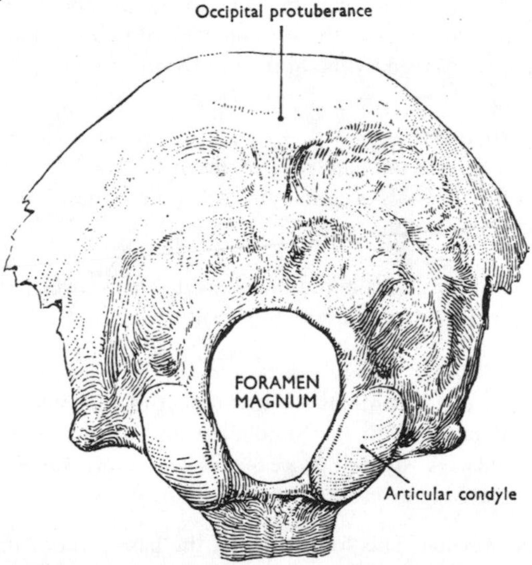

The occipital bone.

The occipital bone

This is the bone which forms the back of the head and part of the base of the skull. It forms immovable joints with the parietal, temporal, and sphenoid bones. Its inner surface is deeply concave and the concavity is occupied by the cerebellum or hind brain and certain large blood vessels. On the outer surface there is a roughened area called the *occipital protuberance* which gives attachment to muscles. The occiput has two articular condyles where it forms a hinge joint with the first bone of the vertebral column, the atlas. Between these condyles there is a large foramen called the *foramen magnum* through which the spinal cord passes.

The sphenoid bone

This bone is in the shape of a bat with its wings outstretched, and it occupies the middle portion of the base of the skull. The 'wings' extend outwards to the sides of the cranium articulating with the temporal, parietal and frontal bones. On the superior surface of the 'body of the bat' there is a little saddle-shaped depression called the *sella turcica* or hypophyseal fossa in which the *pituitary gland* rests. The body of the bone contains some fairly large sinuses, which are lined by ciliated mucous membrane and are in communication with the nasal cavities, and foramina for the passage of blood vessels and nerves.

Figure 39

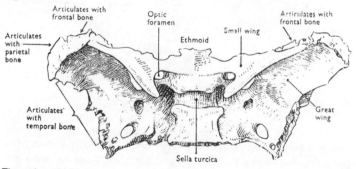

The sphenoid bone.

The ethmoid bone.

The ethmoid bone occupies the anterior part of the base of the skull and helps to form the orbital cavity, the nasal septum and the lateral walls of the nasal cavity. On each side it presents two projections into the nasal cavities called the *upper* and *middle conchae* or *turbinated processes*. It is a very delicate bone containing many air sinuses which have the same characteristics as those of the sphenoid bone. It has a horizontal flattened part called the *cribriform plate* which forms the roof of the nasal cavities and has numerous small foramina through which nerve fibres of the *olfactory nerve* (the nerve of the sense of smell) pass upwards from the nasal cavities to the brain. There is also a very fine *perpendicular plate* of bone which acts as the upper part of the *nasal septum.*

Figure 40

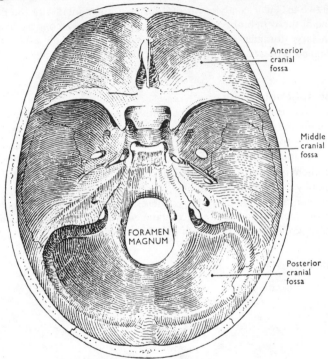

The base of the skull showing the fossae.

The cranium, formed by the bones which have just been described, acts as a bony protection for the brain and is divided for descriptive purposes, into two parts. The *base* and the *vault.* The base is the part on which the brain rests and the vault the part which surrounds and covers it. The base of the cranium is divided into the *anterior, middle* and *posterior cranial fossae* which are deeply indented by the parts of the brain which rest on them and are perforated by many foramina for the passage of blood vessels and nerves passing to and from the brain.

The inner surfaces of all the bones of the cranium are markedly grooved by blood vessels with which they are in contact.

THE FACE
There are 14 bones which form the skeleton of the face but for

Figure 41

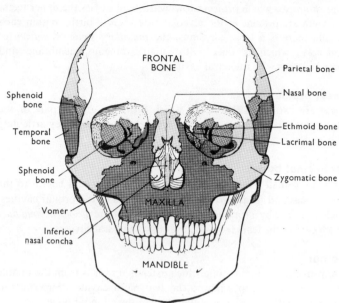

The bones of the skull—Anterior view.

completeness, the frontal bone which has already been discussed
should be added.

Zygomatic or cheek bones	2
Maxillae	2
Nasal bones	2
Lachrymal bones	2
Vomer	1
Palatine bones	2
Inferior conchae or turbinated bones	2
Mandible	1

Zygomatic or cheek bone
Each zygomatic or cheek bone forms the prominence of the cheek
and part of the floor and lateral walls of the orbital cavity. It
articulates with the zygomatic process of the temporal bone to form
the zygomatic arch.

Maxillae or upper jaw bones
The two maxillae or maxillary bones form the upper jaw, the anterior
part of the hard palate or roof of the mouth, the lateral walls of the
nasal cavities and part of the floors of the orbital cavities. The two bones

unite to form one before birth. Together they present the *alveolar ridge* or *process* which projects downwards and carries the upper teeth. The teeth are present in the alveolar ridge before birth. Within each maxilla there is a large air sinus, the *maxillary sinus* or *antrum of Highmore*, which is lined with ciliated mucous membrane and communicates with the nasal cavities.

Nasal bones

These are two small flat bones which form the greater part of the lateral and superior surfaces of the bridge of the nose. They articulate with each other medially.

Lachrymal bones

These two small bones are in a position posterior and lateral to the nasal bones and form part of the medial walls of the orbital cavities. Each is pierced by a foramen for the passage of the *nasolachrymal duct* which carries the tears from the eyes into the nasal cavities.

Vomer

The vomer is a thin flat bone which extends upwards from the middle of the hard palate to separate the two nasal cavities. Superiorly it articulates with the perpendicular plate of the ethmoid bone.

Palatine bones

These are two L-shaped bones. The horizontal parts unite to form the posterior part of the hard palate and the perpendicular parts project upwards to help to form the lateral walls of the nasal cavities. At their upper extremities they form part of the orbital cavities.

Inferior conchae or turbinated bones

Each concha is a long scroll-shaped bone which forms part of the lateral wall of the nasal cavity and projects into it below the middle concha of the ethmoid bone.

Mandible

This is the strongest bone of the face and is the only movable bone of the skull. The mandible consists of two main parts.

A Curved Body. On the superior surface of the curved body is the *alveolar ridge* containing the lower teeth. The anterior surface is pierced by two foramina for the passage of the inferior dental nerves, one on each side.

Two Rami. One ramus projects upwards almost at right angles to each end of the body. At its upper end each ramus divides into two processes. The *condyloid process* articulates with the temporal bone to form the *temporo-mandibular* joint and the *coronoid process* gives attachment to muscles and ligaments.

The points where the rami join the body are called the *angles* of the jaw.

Figure 42

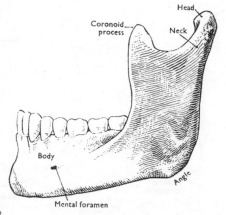

The mandible.

Hyoid bone

This is an isolated horse-shoe-shaped bone lying in the soft tissues of the neck just above the *larynx* or voice box and below the mandible. It does not articulate with any other bone but is attached to the styloid process of the temporal bone by ligaments. It gives attachment to the base of the tongue.

THE SINUSES

It will have been noted that many of the bones which make up the skeleton of the skull contain sinuses. All of these sinuses are in communication with the upper air passages and are lined with ciliated mucous membrane. They serve two important purposes:

1. give resonance to the voice

2. lighten the bones of the face and cranium thus making it easier for the head to balance on top of the vertebral column.

FONTANELLES OF THE SKULL

It has already been stated that the joints between the bones of the skull are called sutures. When ossification of the bones of the skull is complete these sutures are very closely knit, however, at birth ossification is not complete and there are two very marked membranous parts which are called the *anterior and posterior fontanelles*.

Anterior fontanelle

This is diamond shaped and is situated at the junction of the frontal, coronal and sagittal sutures. It is the largest fontanelle and is not fully ossified until the child is 12 to 18 months old.

Figure 43

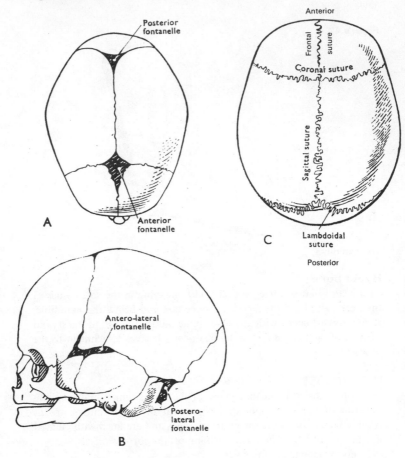

The surface of the skull showing fontanelles and sutures.

(A) Fontanelles as seen from above.
(B) Fontanelles as seen from the side.
(C) Sutures as seen from above.

Posterior fontanelle

This is a smaller triangular-shaped membranous part at the junction of the sagittal and lambdoidal sutures and is usually ossified two to three months after birth.

THE VERTEBRAL COLUMN

The vertebral column consists of 24 separate and movable irregular bones plus the *sacrum* consisting of five fused bones, and the *coccyx* consisting of four fused bones. The 24 separate bones are divided into three groups:

cervical   7
thoracic   12
lumbar   5

All the movable vertebrae have certain characteristics in common but each group has its own special distinguishing features.

Figure 44

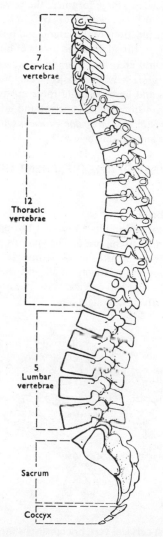

7
Cervical
vertebrae

12
Thoracic
vertebrae

5
Lumbar
vertebrae

Sacrum

Coccyx

General view of the formation of the vertebral column.

CHARACTERISTICS OF A TYPICAL VERTEBRA

A body

Each vertebra has a cylindrical-shaped body which is situated anteriorly. The size of the body varies with the situation of the vertebra, being smaller in the cervical region and becoming larger towards the lumbar end of the column.

A neural arch

The neural arch encloses a large foramen, the *vertebral foramen*, which contains the spinal cord. The *pedicles* or roots of the neural arch project backwards from the junctions of the lateral and posterior aspects of the body; and the *laminae,* broad flattened plates of bone, project medially to complete the neural arch. In the midline, where the laminae meet, there is a *spinous process* which projects backwards. At the points where the pedicles and laminae unite there are two lateral processes called *transverse processes.* On the superior and inferior surfaces of the neural arch, there are two *articular processes* for articulation with the vertebra above and the one below.

SPECIAL FEATURES OF THE DIFFERENT GROUPS OF VERTEBRAE

Cervical vertebrae

Each transverse process exhibits a foramen through which a vertebral artery passes upwards to the brain. The spinous processes are forked or bifid at the ends giving attachment of muscles and ligaments. The bodies of this group are relatively small and the vertebral foramina relatively large. The first two cervical vertebrae are atypical and therefore must be described separately.

Figure 45

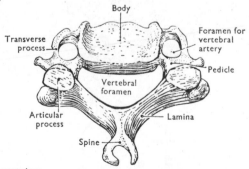

A cervical vertebra.

The Atlas. The first cervical vertebra consists simply of a ring of bone with two short transverse processes. The ring is divided into two parts.

Figure 46

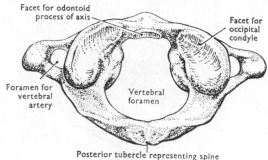

The atlas.

The anterior part is occupied by the odontoid process of the axis, which is held in position by a *transverse ligament*. Thus the odontoid process represents the body of the atlas. The posterior part of the ring is occupied by the spinal cord. On its superior surface the bone has two articular facets which form joints with the condyles of the occipital bone of the skull. The nodding movement of the head takes place at these joints.

The Axis. This is the second cervical vertebra. The body is small and has an upward projecting process called the *odontoid process* or the *dens*. This toothlike process articulates with the first cervical vertebra and the movement at this joint is rotation, i.e., the turning of the head from side to side.

Figure 47

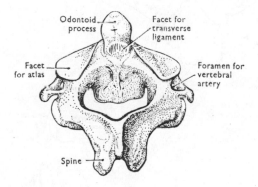

The axis.

Thoracic vertebrae

The spinous processes are long and point downwards so that they partly overlap each other. The thoracic vertebrae articulate with the ribs and so present two half-facets on each lateral surface of the body and one articular surface on each transverse process.

Figure 48

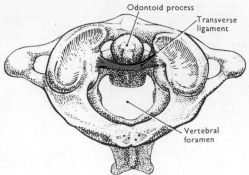

The atlas with the axis in position supported by the transverse ligament.

Lumbar vertebrae

The bodies of the lumbar vertebrae are the largest and the vertebral foramina are the smallest. The spinous processes are short, flat and project straight back.

Figure 49

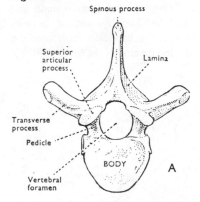

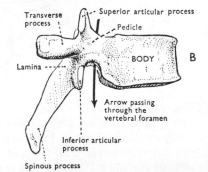

A thoracic vertebra

(A) From above.

(B) From the side.

Figure 50

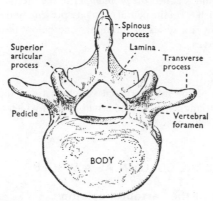

A lumbar vertebra.

The sacrum

This consists of five rudimentary vertebrae fused together to form a wedge-shaped bone with a concave anterior surface. The upper part or the base of the bone articulates with the fifth lumbar vertebra. Laterally it articulates with the two *os coxae* to form the *sacro-iliac joints*, and at its inferior tip it articulates with the *coccyx*. Together the sacrum and the two os coxae form the *pelvis* or the *pelvic girdle*. The bodies of the individual vertebrae can still be distinguished although fused together. The 'body' of the first of these bones protrudes into the

Figure 51

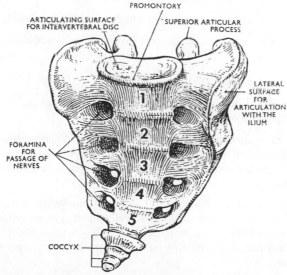

The sacrum and coccyx.

pelvic cavity and is called the *promontory* of the sacrum. The vertebral foramina, which form the neural canal are present, and on each side of the bone there is a series of foramina, one below the other, for the passage of nerves.

The coccyx

This consists of the four terminal vertebrae fused together to form a very small triangular bone the broad base of which articulates with the tip of the sacrum.

THE VERTEBRAL COLUMN AS A WHOLE

The curves of the vertebral column

When viewed laterally it will be seen that the vertebral column is not straight but presents four curves, two of which are described as *primary* and two *secondary*.

When the fetus (the infant before birth) is in the uterus it lies curled up with the vertebral column bent so that the head and the knees are more or less touching. This position shows the primary curvature of the column. After birth this curvature is maintained until, at about three months old, the child can control the movements of his head. This control causes the development of the first secondary curve, the cervical curve. At the age of about 12 to 18 months the child begins to walk thus forming the other secondary curve, the lumbar curve. The primary curves which remain are the thoracic and sacral curves. When standing upright the vertebral column presents four curves.

Thoracic and sacral primary curves which are concave anteriorly.

Cervical and lumbar secondary curves which are convex anteriorly.

Ligaments of the vertebral column

Transverse Ligament. Maintains the odontoid process of the axis in the correct position in relation to the atlas.

Anterior Longitudinal Ligament. Extends the whole length of the column and is situated on the anterior aspect of the bodies of the vertebrae. Its deeper layers are attached to each bone and therefore holds it firmly in position.

Posterior Longitudinal Ligament. Lies within the vertebral canal and extends the whole length of the vertebral column in close contact with the bodies of the bones. It is attached to each bone and plays an important part in maintaining the intervertebral discs in their correct position.

Ligamenta Flava. Connect the laminae of adjacent vertebrae together.

Figure 52

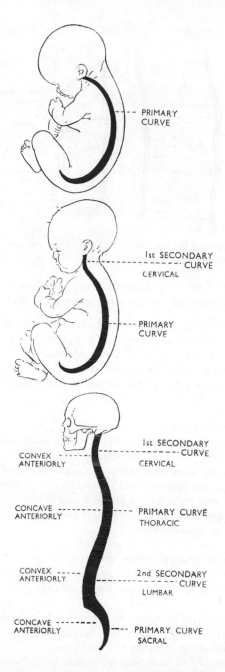

PRIMARY
CURVE

1st SECONDARY
CURVE
CERVICAL

PRIMARY
CURVE

1st SECONDARY
CURVE
CERVICAL

CONVEX
ANTERIORLY

CONCAVE
ANTERIORLY

PRIMARY CURVE
THORACIC

CONVEX
ANTERIORLY

2nd SECONDARY
CURVE
LUMBAR

CONCAVE
ANTERIORLY

PRIMARY CURVE
SACRAL

Diagram illustrating the development of the curves of the spine.

Figure 53

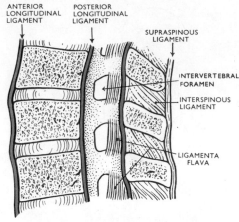

Diagram illustrating the ligaments of the vertebral column.
The vertebrae have been cut longitudinally.

Supraspinous Ligament. Extends from the seventh cervical vertebra to the sacrum connecting the tips of the spinous processes together.

Ligamentum Nuchae. Extends from the occiput to the seventh cervical vertebra connecting the bifid spinous processes together.

The body of each vertebra is firmly connected to its neighbour by a pad of white fibro-cartilage called an *intervertebral disc*. These discs are thinner towards the cervical end and thicken towards the sacral end of the vertebral column.

Movements of the vertebral column

The movements between the individual bones of the vertebral column are very limited. This is necessary because of its important function of protection of the spinal cord. However, the movements of the column as a whole are quite extensive and include *flexion* or bending forward, *extension* or bending backward, *lateral flexion* or bending to the side and *rotation* or turning round. There is more movement in the cervical and lumbar regions than elsewhere.

Functions of the vertebral column

1. Collectively the vertebral foramina form the vertebral canal which provides a strong bony protection for the delicate spinal cord which lies within it.

2. The pedicles of contiguous vertebrae form intervertebral foramina in each side of the vertebral column. It is through these foramina that spinal nerves emerge from the vertebral canal.

3. Because of the numerous individual bones which make it up a certain amount of movement is possible.

4. It supports the skull which is protected from shock by the presence of the intervertebral discs.

5. It forms the axis of the trunk and gives attachment to the ribs, the shoulder girdle and the upper limbs, and the pelvic girdle and lower limbs.

THE THORACIC CAGE

The bones of the thorax or thoracic cage are:

sternum	1
ribs	12 pairs
thoracic vertebrae	12

The sternum or breast bone

This is a flat bone which can be felt just under the skin in the middle of the chest. It is about six inches long, shaped like a dagger and is described in three parts.

The Manubrium. This is the uppermost part, presents two articular facets on its lateral aspects for articulation with the clavicles at the sterno-clavicular joints. The first two pairs of ribs also articulate with the manubrium just below the sterno-clavicular joints.

The Body or Middle Portion. Presents facets for the attachment of five pairs of costal cartilages on its lateral borders.

Figure 54

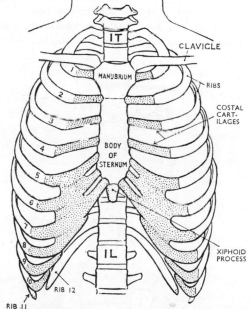

The bones forming the thoracic cage.

Figure 55

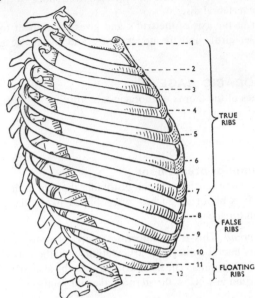

Side view of the sternum and ribs.

The Ensiform or Xiphoid Process. This is the tip of the bone. In some cases this part of the bone is never completely ossified.

The manubrium and body have notches on their lateral aspects which give attachment to the costal cartilages of the ribs. The ensiform process

Figure 56

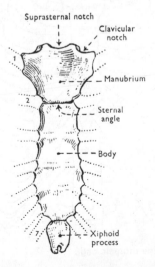

The sternum.

gives attachment to the diaphragm and muscles of the anterior
abdominal wall.

The ribs

There are 12 pairs of ribs which form the bony lateral walls of the
thoracic cage. The first seven pairs are described as *true ribs* because
their anterior ends are attached directly to the sternum by costal
cartilages. The remaining five pairs are called *false ribs*. The eighth,
ninth and tenth pairs are attached by their costal cartilages to the
costal cartilage immediately above so that they are only indirectly
attached to the sternum. The last two pairs have no anterior attachment
and are therefore called *floating ribs*. All 12 pairs of ribs articulate
posteriorly with the thoracic vertebrae.

Figure 57

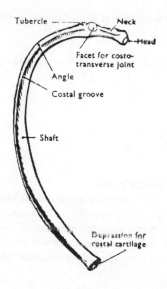

A rib.

Structure of a Rib. A rib is a flat curved bone having a head, neck,
tubercle, angle, sternal end, anterior and posterior surface and a superior
and inferior border.

The head articulates posteriorly with the bodies of two adjacent
vertebrae. The neck is a constricted portion immediately anterior to the
head and between the head and the tubercle. The tubercle articulates
with the transverse process of a thoracic vertebra. The angle is the point
at which the bone bends. The sternal end is attached to the sternum
by a costal cartilage, i.e., a band of hyaline cartilage. The superior

border is rounded and smooth while the inferior border exhibits a marked groove which is occupied by blood vessels and nerves.

The first rib is different from the others in that its broad surfaces are superior and inferior while its borders are anterior and posterior. It does not move during respiration.

The spaces between the ribs are called *intercostal spaces* which in life are filled with muscles which lift the ribs upwards and outwards, thus increasing the size of the thoracic cage laterally. This movement is part of the mechanism of normal inspiration.

Thoracic vertebrae
The twelve thoracic vertebrae have already been described.

The Appendicular Skeleton

THE SHOULDER GIRDLE AND UPPER LIMB OR EXTREMITY
Each shoulder girdle consists of the following bones:

clavicle	1
scapula	1

Each upper extremity consists of the following bones:

humerus	1
radius	1
ulna	1
carpal bones	8
metacarpal bones	5
phalanges	14

The clavicle or collar bone
The clavicle is a long bone which has a double curve. At its rounded medial extremity it articulates with the manubrium of the sternum at the sterno-clavicular joint and at its flattened lateral extremity it forms the acromio-clavicular joint with the acromion process of the scapula. The shaft of the bone is roughened for the attachment of muscles.

Figure 58

Sternal
extremity

Acromial
extremity

A clavicle.

The scapula or shoulder blade
The scapula is a flat triangular-shaped bone which lies on the chest

wall posterior to the ribs and separated from them by muscles. It is described as having:

three borders,
three angles
three fossae.

The three borders of the bone form its outer extremities and meet at its three angles. The borders are called medial, superior and lateral. The angle where the medial and superior borders meet is called the superior angle and where the medial and lateral meet, the inferior angle. At the point where the superior and lateral borders meet there is the lateral angle which presents a shallow articular surface called the

Figure 59

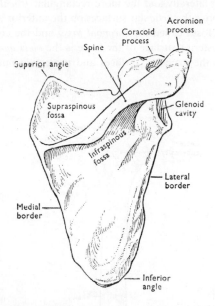

A scapula.

glenoid cavity which, together with the head of the humerus, forms the shoulder joint. Projecting laterally from the superior border of the bone is the *coracoid process* which gives attachment to muscles and ligaments. The posterior surface of the bone is divided into two fossae by a *spine*, these are the *supraspinous fossa* and the *infraspinous fossa*. The spinous process projects beyond the lateral angle of the bone as the *acromion process* which overhangs the shoulder joint and articulates with the lateral extremity of the clavicle at the acromio-clavicular joint. On the anterior surface of the bone there is the *subscapular fossa*. These fossae and borders are grooved and ridged for the attachment of muscles.

The humerus

This is a long bone and is the bone of the upper arm. It consists of a proximal end or head, neck, shaft and distal end.

The head is smooth and rounded and takes part in the formation of the shoulder joint where it articulates with the glenoid cavity of the scapula.

The neck is a slightly constricted part immediately distal to the head. Between the neck and the shaft there are two roughened projections of bone, the *greater and lesser tuberosities* and between them there is a deep groove, the *bicipital groove* which houses one of the tendons of the biceps muscle.

At its proximal end the shaft is cylindrical in shape but becomes flattened on its anterior and posterior surfaces towards the distal end.

The distal end of the bone presents two articular surfaces, the rounded *capitellum* lying laterally and the more rectangular *trochlea* medially. Immediately above the articular surfaces on the anterior aspect of the bone there is a fossa called the *coronoid fossa* and the corresponding fossa on the posterior surface of the bone is the *olecranon fossa*. In a position above the articular surfaces and projecting outwards from

Figure 60

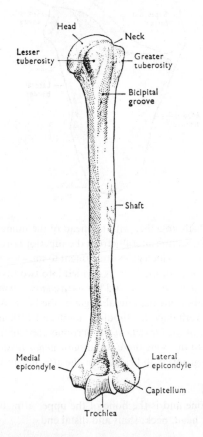

The humerus.

them, one on each side, there are the *medial and lateral epicondyles.* These epicondyles are very close to the surface and therefore can be easily identified by the student.

The ulna and radius

These are the two bones of the forearm. The ulna is medial to the radius and when the arm is in the anatomical position the two bones are parallel. In life there is an interosseous membrane between the bones.

The ulna

This is a long bone consisting of a proximal end, a shaft and a distal end or head. The proximal end of the bone presents two articular fossae and two processes. The *trochlear notch* is a semilunar notch which articulates with the trochlea of the humerus. The proximal extremity of this notch is called the *olecranon process* which forms the point of the elbow and fits into the olecranon fossa of the humerus when the arm is straight. At the distal end of the trochlear notch there is another process, the *coronoid process* which in turn fits into the coronoid fossa of the humerus when the arm is bent. On the lateral surface of the proximal end of the bone there is the other articular surface, the *radial notch.* It is with this notch that the head of the radius articulates to form the *superior radio-ulnar joint.*

The shaft of the bone is triangular in shape and is roughened for the attachment of muscles.

The distal end or head of the ulna is separated from the wrist joint by a pad of white fibro-cartilage. On the lateral aspect there is a smooth surface for articulation with the radius at the *inferior radio-ulnar joint.* The head presents a *styloid process* projecting from its posterior aspect which gives attachment to ligaments.

The radius

The radius is the lateral bone of the forearm and presents a head, neck, tuberosity, shaft and distal extremity.

The head is disc-shaped and flat on top with a very shallow depression which articulates with the capitellum of the humerus. The circumference of the head articulates with the radial notch of the ulna to form the superior radio-ulnar joint.

The neck is a constricted portion of the bone which joins the head to the shaft.

Medially, at the upper end of the shaft, there is the radial tuberosity which gives attachment to muscles. The shaft is roughly triangular in shape being smooth on its lateral surface and more angular and roughened medially.

The distal end of the bone is expanded. It articulates with the carpal bones to form the wrist joint and with the ulna to form the inferior radio-ulnar joint. On its lateral aspect it presents a styloid process for the attachment of ligaments and muscles.

Figure 61

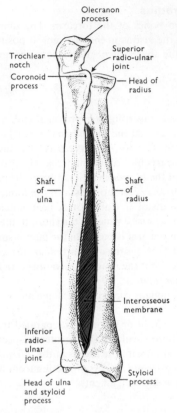

The radius and ulna joined by the interosseous membrane.

The carpal bones or the bones of the wrist.

The carpal bones are eight in number and are arranged in two rows of four. Their names, from without inwards, are enumerated below.

Proximal row: scaphoid, lunate, triquetral, pisiform.

Distal row: trapezium, trapezoid, capitate, hamate.

These bones are all closely fitted together and held in position by ligaments which allow for a certain amount of movement between them. They form joints with each other, the proximal row is associated with the wrist joint and the distal row form joints with the metacarpal bones. They give attachment to the short muscles of the hand which move the fingers and to the muscles which move the wrist joint.

Figure 62

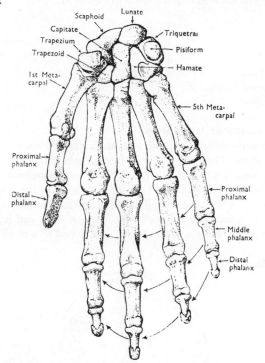

The bones of the wrist, hand and fingers.

The metacarpal bones or the bones of the hand

These are five in number and they form the structure of the palm of the hand. They are not given names but are numbered from the thumb side inwards. They are long slender bones, the proximal ends of which articulate with the carpal bones and the distal ends articulate with the phalanges.

The phalanges or bones of the fingers

There are 14 phalanges which are arranged so that there are three in each finger and two in the thumb. The proximal phalanx of each finger is the largest and articulates with the corresponding metacarpal bone at one end and with the middle phalanx at the other end. The distal phalanx is the smallest and forms the tip of the finger. The thumb which is the shortest of the digits has only two phalanges.

THE PELVIC GIRDLE AND LOWER LIMB OR EXTREMITY

The bones which make up the pelvic girdle or pelvis are:

os coxae	2
sacrum	1

The bones which make up the lower extremity are:

femur	1
tibia	1
fibula	1
patella	1
tarsal bones	7
metatarsal bones	5
phalanges	14

The os coxae or hip bones

Each hip bone consists of three separate bones, the *ilium*, the *ischium* and the *pubis*. These three bones fuse together to form one large irregular bone. On its outer surface it has a deep depression called the *acetabulum* which, with the almost spherical head of the femur forms the hip joint. Above and behind the acetabulum there is a large notch in the bone called the *great sciatic notch* and in front and below there is the *obturator foramen* which, in life, is occluded by a membrane.

Figure 63

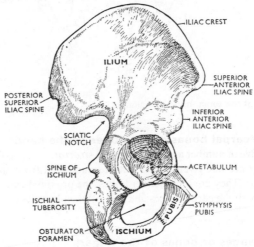

The hip bone.

The upper flattened part of the bone, called the *ilium*, presents the *iliac crest* the anterior point of which is called the *anterior superior iliac spine*.

The *pubis* is the anterior part of the bone and articulates with the pubis of the other hip bone at a slightly movable joint, the *symphysis pubis*.

The *ischium* is the inferior and posterior part of the bone, the most dependent part of which is the *ischial tuberosity*.

The union of the three parts takes place in the acetabulum.

The external surfaces of the innominate bones is markedly ridged for the attachment of muscles.

The pelvis

The pelvis is formed by the os coxae which articulates anteriorly at the symphysis pubis and posteriorly with the sacrum to form the sacro-iliac joints. It is divided into a *false* and a *true pelvis*. Looking down on the pelvis from above it can be seen that there is a ridge of bone which projects into the cavity all the way round. Posteriorly this ridge is

Figure 64

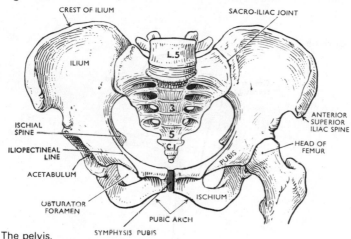

The pelvis.

formed by the promontary of the sacrum and laterally and anteriorly by the *ilio-pectineal line* of the hip bones. The part of the pelvis above this ridge is the false pelvis and the part below, the true pelvis.

DIFFERENCES BETWEEN THE MALE AND FEMALE PELVIS

There are distinct differences between the pelves of the male and the female. The female pelvis is shaped to allow for the passage of the

Figure 65

Diagram illustrating the differences between the male and female pelvis.

fetus during childbirth. The fundamental differences can be summarised in the following manner.

Female	**Male**
Bones. Lighter and smaller.	Heavier and larger.
Cavity. Shallow and round.	Deep and funnel-shaped.
Sacrum. More concave anteriorly making the true pelvis roomier.	Less concave making the true pelvis narrower at the outlet.
Pubic Arch.	
The angle made by the two pubic bones at the symphysis pubis is wide.	The angle of the pubic arch is narrow.

Figure 66

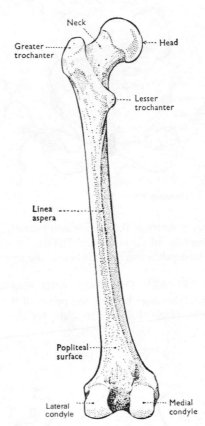

A femur, posterior aspect.

The femur or thigh bone

The femur is the longest and strongest bone of the body and can be described as having two extremities and a shaft.

The proximal extremity consists of a head, neck and *greater and lesser trochanters*. The head is almost spherical in shape and fits into the acetabulum of the hip bone to form the *hip joint*. In the centre of the head there is a small depression for the attachment of the *ligament of the head of the femur* which extends from the acetabulum to the femur conveying a blood vessel which supplies blood to an area of the head of the bone. The neck of the bone extends outwards and slightly downwards from the head to the shaft. The greater and lesser trochanters are two large eminences situated at the junction of the neck and the shaft, with the *intertrochanteric crest* between them on the posterior surface of the bone. These two trochanters and the intertrochanteric crest give attachment to muscles which move the hip joint.

The shaft of the bone is smooth and rounded on its anterior surface but on its posterior surface it has a roughened ridge extending the length of the shaft, called the *linea aspera* which gives attachment to muscles. The shaft of the bone is slightly convex anteriorly and becomes broader towards its distal end. The posterior surface of the lower third forms a flat triangular area called the *popliteal surface*.

The distal extremity presents two articular *condyles* which take part in the formation of the knee joint. Between the condyles there is a deep depression called the *intercondylar notch*. The anterior aspect of the lower end of the bone has an articular surface for the patella called the *patellar surface*.

The tibia or shin bone

The tibia is the medial of the two bones of the leg. It is a long bone and presents two extremities and a shaft.

The proximal extremity is broad and flat and presents two *condyles* for articulation with the condyles of the femur at the knee joint. Between the condyles there is a ridge called the *intercondylar eminence*. Distal to the articular surfaces on the anterior aspect of the bone there is the *tubercle* of the tibia which gives attachment to muscles. The lateral condyle presents an *articular facet* on its inferior surface for articulation with the head of the *fibula* at the *superior tibio-fibular joint*.

The shaft of the bone is roughly triangular in shape and is described as having a medial, lateral and posterior surface. The *crest* of the tibia can be felt very close to the surface on the anterior aspect of the leg.

The distal extremity of the tibia is smooth and flat where it forms the *ankle joint* with the *talus*. On the medial aspect there is a downward projecting process called the *medial malleolus*. The lateral aspect of this process also takes part in the formation of the ankle joint. The lateral aspect of the distal end of the bone articulates with the fibula at the inferior *tibio-fibular* joint.

The fibula

The fibula is a long slender bone which is lateral to the tibia in the leg.

The head or upper extremity articulates with the lateral condyle of the tibia and the lower extremity articulates with the lower extremity of the tibia but projects beyond it to form the *lateral malleolus* and take part in the formation of the ankle joint. The shaft of the bone is ridged for the attachment of muscles. Between the shafts of the tibia and fibula there is an *interosseous membrane*.

Figure 67

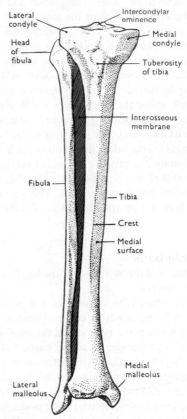

The tibia and fibula showing the interosseous membrane.

The patella or knee cap

This is a *sesamoid* bone associated with the knee joint. It is roughly triangular in shape and lies with the apex pointing downwards. Its anterior surface is in the *patellar tendon* and its posterior surface articulates with the patellar surface of the femur at the knee joint.

The tarsal or ankle bones

The tarsal bones are seven in number.

The *talus* which articulates with the tibia and fibula at the ankle joint.

Figure 68

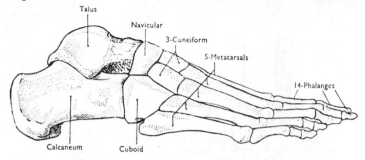

The bones of the foot.

The *calcaneum* or heel bone which is markedly roughened for the attachment of muscles which move the ankle joint.

The *navicular* which is situated on the medial aspect of the foot distal to·the talus.

The *medial, intermediate* and *lateral cuneiforms* and the *cuboid* form a row of bones from within outwards which articulate with the other three tarsal bones proximally and with the five metatarsal bones distally.

The metatarsal bones

These are five in number and form the greater part of the dorsum of the foot. At their proximal ends they articulate with the tarsal bones and at their distal ends, with the phalanges. They do not have individual names but are numbered from within outwards.

The phalanges

There are 14 phalanges in each foot which are arranged in a similar manner to those in the fingers, i.e., two to the great toe and three to each of the other toes.

ARCHES OF THE FOOT

The arrangement of the bones of the foot is such that the foot is not a rigid structure. This point is well seen by comparing a normal foot with a 'flat' foot. The bones have a bridge-like arrangement and are supported by muscles and ligaments so that four arches are formed:

  medial longitudinal arch
  lateral longitudinal arch
  two transverse arches.

Medial longitudinal arch

This is the highest of the arches and is formed by the calcaneum, navicular, three cuneiform and first three metatarsal bones. Only the calcaneum and the metatarsal bones should touch the ground.

Figure 69

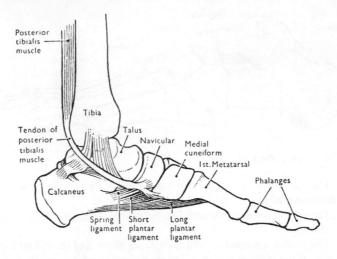

Diagram illustrating the ligaments which help to support the arches of the foot.

Lateral longitudinal arch

The lateral arch is much less marked than its medial counterpart and has only four bony components, the calcaneum, cuboid and two lateral metatarsal bones. Again only the calcaneum and metatarsal bones should touch the ground.

Transverse arches

These arches run across the foot and can be more easily seen by examining the skeleton than the live model. They are most marked at the level of the three cuneiform and cuboid bones.

MUSCLES AND LIGAMENTS WHICH SUPPORT THE ARCHES OF THE FOOT

As there are movable joints between all the bones of the foot very strong muscles and ligaments are necessary to maintain the strength, resilience and stability of the foot during walking, running and jumping.

Posterior tibialis muscle

This is the most important muscular support of the medial longitudinal arch. It lies on the posterior aspect of the lower leg, originates from the middle third of the tibia and fibula and its tendon passes behind the medial malleolus to be inserted into the navicular, cuneiform, cuboid and metatarsal bones. It can be seen, therefore, that it acts as a sling or 'suspension apparatus' for the arch.

Short muscles of the foot

This group of muscles is mainly concerned with the maintenance of the lateral longitudinal and transverse arches. They make up the fleshy part of the sole of the foot.

Plantar calcaneonavicular ligament or 'spring' ligament.

This is a very strong thick ligament stretching from the calcaneum to the navicular bone. It plays an important part in supporting the medial longitudinal arch.

Plantar ligaments and interosseous membranes

These structures support the lateral and transverse arches.

3. The Joints or Articulations

A joint is the site at which any two or more bones come together. Some joints have no movement, some only slight movement and some are freely movable. Before going on to the classification of joints the movements which can occur should be considered.

Flexion or bending, usually forward but occasionally backward as in the case of the knee joint.

Extension means straightening or bending backward.

Abduction is movement away from the midline of the body.

Adduction is movement towards the midline of the body.

Rotation is movement round the long axis of a part.

Pronation, that is, turning the palm of the hand down.

Supination means turning the palm of the hand up.

Circumduction is the combination of flexion, extension, abduction and adduction.

Inversion is the turning of the sole of the foot towards the midline.

Eversion consists of the opposite movement to inversion, that is, turning the sole of the foot outwards.

Classification of Joints
FIBROUS OR FIXED JOINTS
In this type of joint there is no movement between the bones concerned.

Figure 70

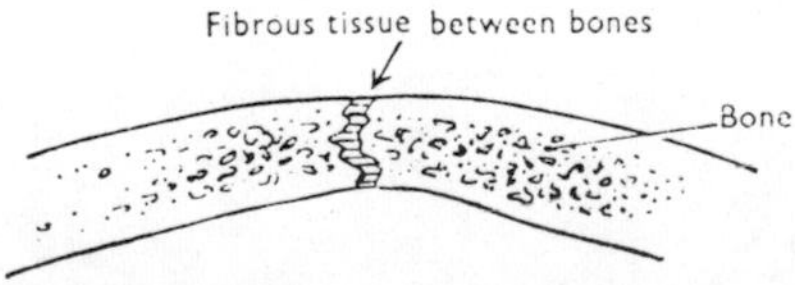

Fibrous tissue between bones

Bone

Diagram illustrating a fibrous joint.

As the name suggests, there is *fibrous tissue* between the ends of the bones. Examples of this type include the sutures of the skull and the teeth in their sockets.

CARTILAGINOUS OR SLIGHTLY MOVABLE JOINTS

In this case there is a pad of *white fibrocartilage* between the ends of the bones taking part in the joint, which allows for very slight movement. Movement is only possible because of compression of the pad of cartilage. Examples of cartilaginous joints include the symphysis pubis and the joints between the bodies of the vertebrae.

Figure 71

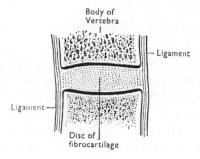

Diagrammatic illustration of a cartilaginous joint.

SYNOVIAL OR FREELY MOVABLE JOINTS

Synovial joints are characterised by the presence of *synovial membrane*.

A considerable amount of movement is permitted at all synovial joints and the types are subdivided according to the movements possible.

Ball and Socket. These are the most freely movable of all joints. The movements possible are flexion, extension, abduction, adduction, rotation and circumduction. Examples are the shoulder and the hip joints.

Hinge Joints. This type permits movement in one plane only. The movements are flexion and extension and examples include the elbow, knee, ankle, the joints between the atlas and the occipital bone and the interphalangeal joints of the fingers and toes.

Gliding Joints. In this case the articular surfaces glide over each other. Examples of this type include the sterno-clavicular joints, the acromio-clavicular joints and the joints between the carpal bones and between the tarsal bones.

Pivot Joints. These joints allow movement round one axis only, that is, a rotatory movement. The classical examples are the superior and inferior radio-ulnar joints and between the atlas and the odontoid process of the axis.

Condyloid and Saddle Joints. In these joints the movements take place round two axes thus permitting the movements of flexion, extension, abduction, adduction and the combination of these is circumduction. Examples include the wrist joint, the temporomandibular joint, metacarpo-phalangeal and metatarso-phalangeal joints.

CHARACTERISTICS OF A SYNOVIAL JOINT

All synovial joints have certain characteristics in common and when describing them it is convenient to do so under the following headings.

1. The *type* of synovial joint.

2. The names of the *parts of the bones* concerned in its formation.

3. *Articular or hyaline cartilage.* The parts of the bones forming the joint are always covered with hyaline cartilage. This tissue is strong enough to bear the weight of the body, as is necessary in the ankle joint, and at the same time provides a smooth surface for articulation.

4. *Capsular ligament.* The joint is surrounded and enclosed by a sleeve of fibrous tissue which joints the bones together. It is sufficiently loose to allow for the range of movement of which the joint is capable.

5. *Intra-capsular structures.* Some joints have structures within the joint capsule, but outside the synovial membrane, which are necessary to ensure their stability.

6. *Synovial membrane.* This is composed of secretory epithelial cells which secrete a thick sticky fluid of the consistency of white of egg, which acts as a lubricant to the joint, provides nutrient materials for the structures within the joint cavity and helps to maintain the stability of the joint. It prevents the ends of the bones from being separated as does a little water between two glass surfaces.

Synovial membrane is found in the joints in well-defined situations.

(*a*) Lining the capsular ligament.

(*b*) Covering those parts of the bones within the joint capsule which are not covered with hyaline cartilage.

(*c*) Covering all intra-capsular structures.

Little sacs of synovial fluid or *bursae* are found in some joints. Their position is such that they act as cushions to prevent friction between a bone and a ligament or tendon.

7. *Extra-capsular structures.* Most joints have ligaments outside the capsular ligament which strengthen and lend stability to the joint.

8. *Muscles and movements.* The contraction of muscles is responsible for producing the movements at individual joints, therefore, they are best considered together.

9. *Nerve supply*. A useful generalisation is that any nerves which cross a joint supply the muscles which move it and the structures which form the joint.

Only the major joints of the body, that is, the joints of the limbs, will be described in detail.

The Shoulder Joint

The shoulder joint is a synovial joint of the *ball and socket type*.

The parts of the bones forming the joint are the glenoid cavity of the scapula and the head of the humerus.

Articular cartilage lines the glenoid cavity of the scapula and covers the head of the humerus.

The capsular ligament which surrounds and encloses the joint is very loose inferiorly to allow for the free movement normally possible at this joint.

Intra-capsular Structures. The glenoid cavity is deepened by a rim of fibrocartilage called the *glenoid labrum*. By deepening the cavity in this way the joint is made more stable but the range of movement is not reduced. The *long head of the biceps muscle*, which lies in the bicipital groove of the humerus, extends through the joint cavity and is attached to the rim of the glenoid cavity. It has an important stabilising effect on the joint.

Synovial membrane lines the capsular ligament and covers that part of the bone within the joint capsule not covered with hyaline cartilage. It forms a sleeve round the part of the long head of the biceps muscles which is within the capsular ligament and it covers the glenoid labrum on both sides. Synovial fluid is secreted by the membrane and lubricates the joint.

Extra-capsular Structures. The most important strengthening ligament is the *coraco-humeral ligament* which extends from the coracoid process of the scapula to the humerus.

MUSCLES AND MOVEMENTS

Flexion: coraco-brachialis, anterior fibres of deltoid and pectoralis major.

Extension: teres major, latissimus dorsi and posterior fibres of deltoid.

Abduction: mainly the central fibres of deltoid.

Adduction: combined action of flexors and extensors.

Medial Rotation: pectoralis major and latissimus dorsi.

Lateral Rotation: mainly posterior fibres of deltoid.

Circumduction: flexors, extensors, abductors and adductors.

Figure 72

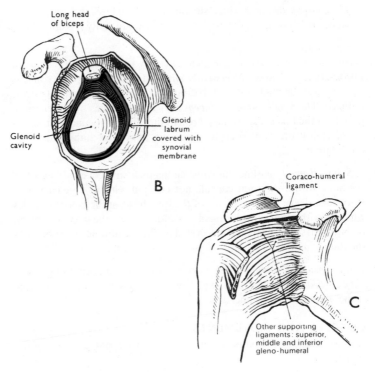

Diagrammatic illustrations of the shoulder joint.

(A) Showing the position of
    the synovial membrane.

(B) Showing the position of the
    glenoid labrum.

(C) Showing the supporting
    ligaments.

Pectoralis major

This is a broad thick almost fan-shaped muscle which lies on the anterior thoracic wall. The fibres originate from the middle third of the clavicle and from the sternum and are inserted into the lateral lip of the bicipital groove of the humerus.

This muscle draws the arm forward and towards the body, that is, flexes and adducts (see Fig. 75).

Deltoid muscle

The deltoid muscle is triangular in shape and lies directly over the shoulder joint. The fibres originate from the clavicle, the acromion process and spine of the scapula and radiate over the shoulder joint to be inserted into the deltoid tuberosity of the humerus.

This muscle has the ability to act in three separate parts. The anterior fibres assist the pectoralis major in flexing the arm, the middle or main part abducts the arm, and the posterior fibres assist the latissimus dorsi in extending the arm.

Coraco-brachialis muscle

This muscle is situated on the upper medial aspect of the arm. It arises from the coracoid process of the scapula, stretches across anterior to the shoulder joint and is inserted into the middle third of the humerus.

Its action is to assist with the movement of flexion of the arm.

Latissimus dorsi

This is a large triangular shaped muscle which has its origin in the posterior part of the iliac crest and the spinous processes of the lumbar and lower thoracic vertebrae. It passes obliquely upwards across the back to be inserted by a narrow tendon to the floor of the bicipital groove of the humerus, having passed under the arm. Its action is to medially rotate and adduct the humerus. It also assists in extension of the arm at the shoulder joint.

Teres major

This muscle originates from the inferior angle of the scapula and is inserted into the humerus just below the shoulder joint. Its action is to draw the humerus backwards and towards the body.

Elbow Joint

The elbow joint is a synovial joint of the *hinge* type.

The parts of the bones concerned are the trochlea and the capitellum of the humerus which articulate with the trochlear notch of the ulna and the head of the radius respectively.

Articular cartilage covers the articular surfaces of the bones.

The Capsular Ligament. This acts like a sleeve surrounding and enclosing the joint. It extends from just above the articular surfaces of the humerus to just below the articular surfaces of the radius and ulna.

Intra-capsular Structures. There are several pads of fat and bursae situated within the capsule.

Synovial Membrane. This secretes synovial fluid which lubricates the joint. It is found lining the capsular ligament, covering the parts of the bone within the capsule which are not covered by hyaline cartilage and covering the pads of fat.

Extra-capsular Structures. These consist of anterior, posterior, medial and lateral strengthening ligaments.

Figure 73

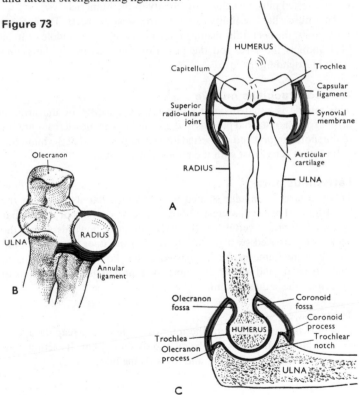

Diagrammatic illustrations of the elbow joint.

(A) Showing the inside of the joint from the front.

(B) Showing the annular ligament.

(C) Showing the inside of the joint from the side.

MUSCLES AND MOVEMENTS

As the elbow joint is a hinge joint there are only two possible movements.

Flexion: biceps and brachialis.

Extension: triceps.

Biceps muscle

This is the large muscle which lies on the anterior aspect of the upper arm. It is given its name because it rises from two tendons or heads. The short head rises from the coracoid process of the scapula and passes in front of the shoulder joint down to the arm. The long head originates from the rim of the glenoid cavity and passes through the joint cavity and the bicipital groove to the arm. The long head of the muscle is retained in the bicipital groove of the humerus by a *transverse ligament* which stretches across the groove. The two parts of the muscle are inserted by one tendon into the tuberosity of the radius.

The action of this muscle is to supinate the hand, flex the elbow joint and lend stability to the shoulder joint (see Fig. 75).

Brachialis

This muscle lies on the anterior aspect of the upper arm deep to the biceps. It originates from the shaft of the humerus, extends across the elbow joint and is inserted into the ulna just distal to the joint capsule.

Its action is to assist the biceps muscle in flexing the elbow joint.

Triceps muscle

This muscle lies on the posterior aspect of the humerus. As the name suggests it arises from three heads, one from the scapula and two from the posterior surface of the humerus. The insertion of the muscle is by a single tendon to the olecranon process of the ulna.

The most important action of the muscle is to extend the elbow joint.

SUPERIOR AND INFERIOR RADIO-ULNAR JOINTS

Incorporated in the same capsule as the elbow joint is the *superior radio-ulnar joint.* This joint is formed by the rim of the head of the radius rotating in the radial notch of the ulna. The articular surfaces of the bones are covered with hyaline cartilage. The *annular ligament* is a strong extra-capsular ligament which encircles the head of the radius and keeps it in contact with the radial notch of the ulna.

The *inferior radio-ulnar joint* is a synovial pivot joint between the distal end of the radius and the head of the ulna

Muscles and movements

Pronation: pronator teres.

Supination: supinators and biceps.

Pronator teres

This muscle lies obliquely across the front of the forearm. It arises from the medial epicondyle of the humerus and the coronoid process of the ulna, passes obliquely across the forearm to be inserted into the lateral surface of the shaft of the radius.

The action of this muscle is to rotate the radio-ulnar joints, changing the hand from the anatomical to the writing position, that is, pronating the forearm.

Supinator muscle

This muscle lies obliquely across the posterior and lateral aspects of the forearm. Its fibres arise from the lateral epicondyle of the humerus and the upper part of the ulna and are inserted into the lateral surface of the upper third of the radius.

Its action is to rotate the radio-ulnar joints, changing the hand from the writing to the anatomical position.

Wrist Joint

The wrist joint is a synovial joint of the *condyloid type*.

The parts of the bones concerned are the distal end of the radius and the proximal ends of the scaphoid, lunate and triquetral. The distal end of the ulna is separated from the joint cavity by a disc of *white fibro-cartilage*, the under surface of which articulates with the carpal bones of the joint. This disc of fibro-cartilage also separates the inferior radio-ulnar joint from the wrist joint.

Hyaline cartilage covers the articular surfaces of the bones.

Capsular ligament surrounds and encloses the joint.

Intra-capsular structures. There are none of significance in this joint.

Synovial membrane secretes synovial fluid which lubricates the joint. It is found lining the capsular ligament and covering the parts of the bones within the capsule which are not covered by hyaline cartilage.

Extra-capsular structures consist of medial and lateral ligaments and anterior and posterior radiocarpal ligaments.

MUSCLES AND MOVEMENTS

The wrist joint is capable of a wide range of movement.

Flexion: flexor carpi radialis and the flexor carpi ulnaris.

Extension: extensor carpi radialis and the extensor carpi ulnaris.

Abduction: flexor and extensor carpi radialis.

Adduction: flexor and extensor carpi ulnaris.

Figure 74

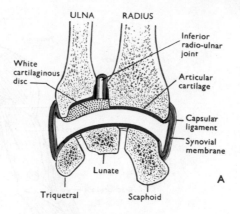

A

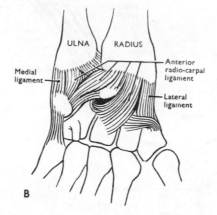

B

Diagrammatic illustrations of the wrist joint.

(A) Showing the structures of the joint from the front.

(B) Showing the supporting ligaments.

Flexor carpi radialis

This muscle originates from the medial epicondyle of the humerus, lies on the anterior surface of the forearm and is inserted into the second and third metacarpal bones.

This is one of the flexor muscles of wrist joint and when it acts with the extensor carpi radialis it abducts the joint (see Fig. 75).

Flexor carpi ulnaris

This muscle lies on the medial aspect of the forearm. It originates from

Figure 75

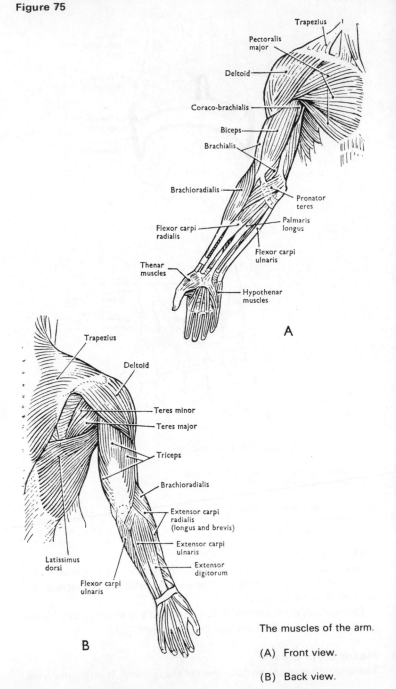

The muscles of the arm.

(A) Front view.

(B) Back view.

the medial epicondyle of the humerus and the upper parts of the ulna and is inserted into the pisiform, the hamate and fifth metacarpal bones.

When contracted with the flexor carpi radialis it flexes the wrist, and when it acts with the extensor carpi ulnaris it adducts the joint.

Extensor carpi radialis longus and brevis

These muscles lie on the posterior aspect of the forearm. The fibres originate from the lateral epicondyle of the humerus and are inserted by a long tendon into the second and third metacarpal bones.

They are associated with extension and abduction of the wrist joint.

Extensor carpi ulnaris

This muscle, like the other extensor muscle, lies on the posterior surface of the forearm. It originates from the lateral epicondyle of the humerus and is inserted into the fifth metacarpal bone.

It is associated with the movements of extension and adduction of the wrist.

Joints of the Hands and Fingers

There is a number of synovial joints between the carpal bones, between the carpal and metacarpal bones, between the metacarpal and proximal phalanges and between the phalanges. The powerful movements which take place at these joints are produced by muscles in the forearms which have tendons extending into the hand. Many of the finer movements are produced by numerous small muscles in the hand.

Hip Joint

The hip joint is a synovial joint of the *ball and socket type.*

The parts of the bones concerned are the cup-shaped acetabulum of the hip bone and the rounded head of the femur.

Articular cartilage lines the acetabulum and covers the head of the femur.

The capsular ligament is a thick strong sleeve of fibrous tissue holding the bones together. It extends from just above the rim of the acetabulum to about half-way along the neck of the femur.

Intra-capsular Structures. The acetabular labrum, like the glenoid labrum of the shoulder joint, is a ring of fibrocartilage attached to the rim of the acetabulum which deepens the cavity, lends stability to the joint but does not interfere with its range of movement. The *ligament of the head of the femur* is a ligament extending from the shallow depression in the middle of the head of the femur to the rim of the lower part of the acetabulum. It conveys a blood vessel to the head of the femur.

Synovial membrane secretes synovial fluid which lubricates the joint. It is found lining the capsular ligament and covering the parts of the bone within the joint cavity which are not covered with articular cartilage. It covers both sides of the acetabular labrum and encases the ligament of the head of the femur.

Figure 76

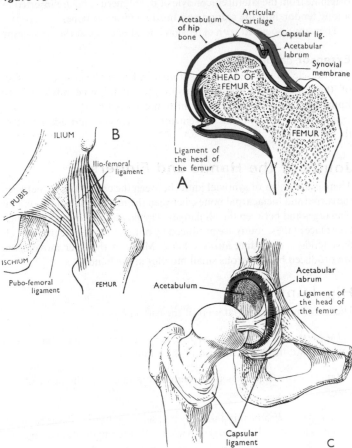

Diagrammatic illustrations of the hip joint.

(A) Showing the structures inside
    the joint.

(B) Showing the supporting
    ligaments.

(C) Showing the acetabular labrum
    and ligament of head of femur.

Extra-capsular Structures. There are three important ligaments surrounding and giving strength to the joint.

1. The ilio-femoral ligament, an inverted Y-shaped ligament lying anteriorly.
2. The ischio-femoral ligament lying posteriorly.
3. The pubo-femoral ligament lying inferiorly.

MUSCLES AND MOVEMENTS

Flexion: rectus femoris, illio-psoas and sartorius.

Extension: gluteus maximus and the hamstrings.

Abduction: gluteus medius and minimus, sartorius and others.

Adduction: adductor group.

Lateral Rotation: mainly gluteal muscles and adductor group.

Medial Rotation: gluteus medius and minimus and others.

Figure 77

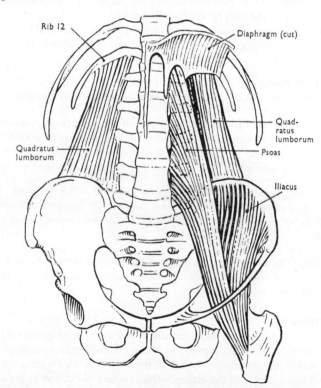

The muscles which form the posterior abdominal wall.

Psoas muscle

This muscle arises from the transverse processes and the bodies of the lumbar vertebrae. It passes across the bones of the false pelvis behind the inguinal ligament to be inserted into the lesser trochanter of the femur.

Its action together with the iliacus is to cause flexion at the hip joint.

Iliacus muscle

This is a triangular shaped muscle which lies in the iliac fossa of the hip bone. It originates from the inner aspect of the iliac crest, passes over the iliac fossa and joins the tendon of the psoas muscle to be inserted into the lesser trochanter of the femur.

The combined muscle, the *illio-psoas* muscle, causes flexion at the hip joint.

Quadriceps femoris

This is a group of four muscles which lie on the front of the thigh. They are the *rectus femoris* and *three vasti muscles*. The rectus femoris originates from the ilium and the three vasti from the upper end of the femur. Together they pass over the front of the knee to be inserted into the tubercle of the tibia by the patellar tendon.

The rectus femoris plays a part in flexing the hip joint and together the whole group acts as a very strong extensor of the knee joint.

Gluteal muscles

This group consists of the gluteus maximus, medius and minimus.

Gluteus maximus forms the fleshy part of the buttock. It originates from the outer surface of the iliac bone and the sacrum and is inserted into the upper end of the femur. It is the muscle chiefly concerned with the maintenance of the erect position by extension of the hip joint.

Gluteus medius lies below gluteus maximus. It originates from the outer surface of the ilium and is inserted into the greater trochanter of the femur. It is associated with abduction and medial rotation of the thigh.

Gluteus minimus is the most deeply situated of the three muscles with its origin and insertion similar to those of gluteus medius. Its action is to abduct and medially rotate the thigh.

Sartorius

This is the longest individual muscle in the body. It originates from the anterior superior iliac spine and passes obliquely across the thigh to be inserted into the medial surface of the upper part of the tibia.

It is associated with the actions of flexion and abduction at the hip joint and flexion at the knee.

Figure 78   The Joints or Articulations   85

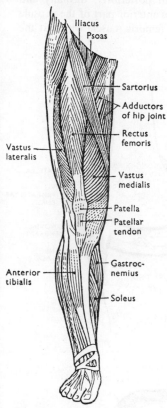

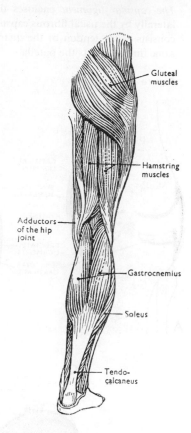

The muscles of the leg.

(A) Viewed from the front.

(B) Viewed from the back.

Adductor group

This group lies on the medial aspect of the thigh. The muscles making up the group all originate from the pubic bone and are inserted into the linea aspera of the femur.

As the name suggests this group adducts the thigh.

Knee Joint

The knee joint is a synovial joint of the *hinge type*.

The parts of the bones concerned are the condyles of the femur, the condyles of the tibia and the posterior surface of the patella.

Articular cartilage covers the articular surfaces of the three bones.

The capsular ligament encloses the joint posteriorly, medially and laterally by the usual fibrous capsule. The anterior part of the capsule consists of the tendon of the quadratus femoris muscle which at the same time supports the patella.

Figure 79

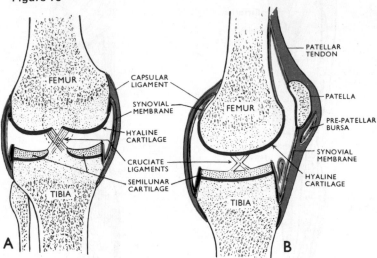

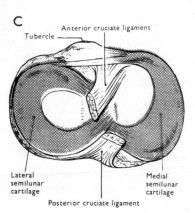

Diagrammatic illustration of the knee joint.

(A) Anterior view showing the intracapsular structures.

(B) Lateral view showing the position of the patella.

(C) View of the superior surface of the tibia showing the semilunar cartilages.

Intra-capsular Structures

1. *Cruciate Ligaments.* There are two cruciate ligaments which cross each other extending from the intercondylar notch of the femur to the intercondylar eminence of the tibia. They have an important stabilising effect on the joint.

2. *Semilunar Cartilages or Menisci.* These are incomplete discs of white fibrocartilage which lie on top of the articular condyles of the tibia. They are wedge-shaped, being thicker at their outer edges. They have a stabilising effect on the joint by preventing lateral displacement of the bones.

3. *Bursae and Pads of Fat.* Numerous bursae and pads of fat are to be found within the capsule of this joint. They are situated so that they prevent friction between a bone and a ligament or tendon.

Synovial Membrane. This membrane lines the capsular ligament and the inner surface of the patellar tendon and covers the parts of the bone within the capsule which are not covered with hyaline cartilage. It covers the cruciate ligaments and the pads of fat. The menisci are not covered with synovial membrane because it is too delicate to bear the weight of the body. Synovial fluid is secreted into the joint cavity.

Extra-capsular Structures. The two most important strengthening ligaments of the knee joint are the medial and lateral ligaments.

MUSCLES AND MOVEMENTS
As this is a hinge joint the movements possible are flexion and extension.

Flexion: gastrocnemius and hamstrings.

Extension: quadratus femoris muscle.

Hamstring muscles
This group of muscles lies on the posterior aspect of the thigh. All the muscles making up the group originate from the ischium and are inserted into the upper end of the tibia. They are the biceps femoris, semimembranosus and semitendinosus muscles.

Their chief action is to flex the knee joint.

Gastrocnemius
This is the muscle which forms the bulk of the calf of the leg. It arises by two heads, one from each condyle of the femur, and passes down behind the tibia to be inserted into the calcaneum by the *tendocalcaneus* or the *tendon of Achilles*.

Quadratus femoris
This group has already been described on page 84.

Contraction of this very strong muscle plantar flexes the foot. It is, therefore, used extensively in walking, running and jumping.

Ankle Joint

The ankle joint is a synovial joint of the *hinge type*.

The parts of the bones concerned are the distal end of the tibia and its malleolus, the malleolus of the fibula and the talus.

Hyaline cartilage covers the articular surfaces of the bones.

The *capsular ligament* surrounds and encloses the joint.

Figure 80

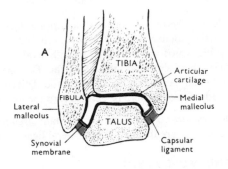

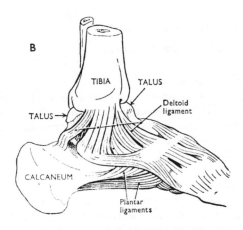

Diagrammatic illustrations of the ankle joint.

(A) Showing the structures within the joint.

(B) Showing the supporting ligaments.

Intra-capsular Structures. There are no structures of significance in this joint.

Synovial membrane is found lining the capsular ligament and covering the parts of the bone within the joint capsule which are not covered with articular cartilage. It secretes synovial fluid into the joint cavity.

Extra-capsular Structures. There are four important ligaments strengthening this joint. They are anterior, posterior and lateral ligaments and a very strong medial ligament called the deltoid ligament.

MUSCLES AND MOVEMENTS
Special names are given to the movements in this case.

Flexion or dorsiflexion: anterior tibialis assisted by the muscles which extend the toes.

Extension or plantarflexion: gastrocnemius and soleus assisted by the muscles which flex the toes.

The movements of *inversion* and *eversion* mentioned previously occur between the tarsal bones and not at the ankle joint.

Anterior tibialis muscle
This muscle originates from the upper end of the tibia and passes down the anterior surface of the lower leg to be inserted into the middle cuneiform bone by a long tendon.

It is associated with dorsiflexion of the foot.

Soleus
This is the other main muscle of the calf of the leg lying immediately deep to the gastrocnemius. It originates from the head and upper part of the fibula. Its tendon joins that of the gastrocnemius so that they have a common insertion into the calcaneum by the tendo-calcaneus.

This muscle is chiefly concerned with the maintenance of the upright position.

Gastrocnemius
This muscle was described previously in relation to the movements of the knee joint.

Joints of the Foot and Toes
There is a number of synovial joints between the bones of the foot and the toes. The strong movements are produced by the muscles in the leg whose tendons cross the ankle. Some of these muscles together with short muscles of the foot support the arches of the foot described previously.

4. The Muscular System

The skeletal muscles described in this chapter are those which are not directly involved in the movements of the joints of the limbs.

The Muscles of Facial Expression and Mastication

There are a large number of muscles which are concerned with alteration of facial expression and with the movements of the lower jaw during mastication. Only the main muscles concerned will be considered here.

Occipitofrontalis muscle

This consists of a posterior muscular part which lies over the occipital bone, an anterior part over the frontal bone and an extensive flat tendon or aponeurosis which stretches over the dome of the skull and joins the two muscular parts together.

The action of this muscle is to raise the eyebrows.

Levator palpebrae superioris

This muscle extends from the posterior part of the orbital cavity to the upper eyelid.

The action of this muscle is to raise the eyelid.

Orbicularis oculi

This muscle surrounds the eye and includes the eyelid and the area immediately round the orbital cavity; it closes the eye and when strongly contracted 'screws up' the eyes.

Buccinator

This is the flat muscle of the cheek. Its action is to draw the cheeks in towards the teeth in chewing and to expel air from the mouth. It is sometimes described as 'the trumpeters muscle'.

Orbicularis oris

This is a circular muscle which surrounds the mouth and blends with the muscles of the cheeks.

Its action is to close the lips and, when strongly contracted, shapes them as for whistling.

Figure 81

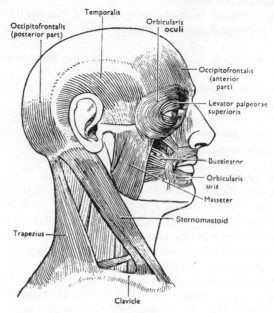

The muscles of the face.

Masseter

This is a broad muscle which extends from the zygomatic arch to the angle of the jaw. Its action is to draw the mandible up to the maxillae when chewing and it is capable of exerting considerable pressure on the food being chewed.

Temporalis

This muscle is shaped like a fan, covering the squamous portion of the temporal bone. It passes under the zygomatic arch to be inserted into the coronoid process of the mandible. Its action is to assist in closing the mouth.

Pterygoid muscles

These muscles extend from the greater wing of the sphenoid bone to the mandible. Its action is to close the mouth and pull the lower jaw forward.

Muscles of the Neck

There are a great many muscles situated in the neck extending from the mandible to the hyoid bone and to the sternum and clavicle, only two of which will be considered here.

Sternomastoid muscles (2)   Trapezius muscles (2).

Sternomastoid muscle

This muscle arises from the manubrium of the sternum and the clavicle and extends upwards to the mastoid process of the temporal bone.

When the muscle on one side contracts it draws the head towards the shoulder.

When both muscles contract together they:

1. flex the cervical vertebrae
2. draw the sternum and clavicle upwards when the head is maintained in a fixed position which occurs mainly in forced respiration.

Trapezius muscle

This muscle covers the shoulder and the upper part of the back of the neck. The upper attachment is to the occipital protuberance, the medial attachment to the transverse processes of the cervical and thoracic vertebrae and the lateral attachment is to the lateral part of the clavicle and to the acromion process of the scapula.

The actions of this muscle include:

1. pulling the head backwards
2. squaring the shoulders and controlling the movements of the scapula when the shoulder joint is in use.

Muscles of the Back

There are six pairs of large muscles in the back in addition to those which form the posterior part of the abdominal wall. The arrangement of these muscles is the same on each side of the vertebral column.

Trapezius	2
Teres major	2
Psoas	2
Sacrospinalis	2
Latissimus dorsi	2
Quadratus lumborum	2

Teres major and latissimus dorsi

These muscles were described previously in relation to the movements of the shoulder joint.

Psoas muscle

(See page 84).

Figure 82

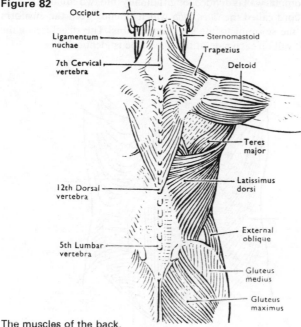

The muscles of the back.

Quadratus lumborum

The origin of this muscle is the posterior part of the crest of the ilium then it passes upwards, parallel and close to the vertebral column to be inserted into the lower rib.

Together the two muscles have the action of fixing the lower rib during respiration and causing backward flexion or extension of the vertebral column. If one muscle contracts it will cause lateral flexion of the lumbar region of the vertebral column.

Sacrospinalis muscle

This is the name given to a group of muscles which lie between the spinous and transverse processes of the vertebrae. It originates from the sacrum and is finally inserted into the occipital bone.

The main action of this group is to produce backward flexion or extension of the vertebral column.

Muscles of the Anterior Abdominal Wall

There are four layers of muscle which make up the anterior abdominal wall.

Rectus abdominis	2
External oblique	2
Internal oblique	2
Transversus abdominis	2

The abdominal wall is divided longitudinally into two equal parts by a tendinous cord called the *linea alba* which extends from the ensiform process of the sternum to the symphysis pubis. The structure of the abdominal wall on each side of the linea alba is identical.

Figure 83

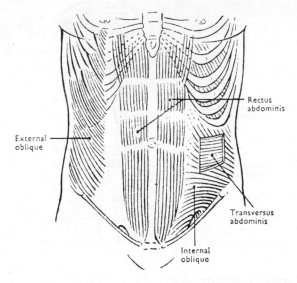

The muscles of the anterior abdominal wall.

Rectus abdominis

This is the most superficial of the four muscles. It is a broad flat muscle originating from the transverse part of the pubic bone then passing upwards to be inserted into the lower ribs and the ensiform process of the stermun. Medially it is closely associated with the linea alba.

External oblique

This muscle extends from the lower ribs *downwards and forward* to be inserted into the iliac crest and, by an aponeurosis, to the linea alba.

Internal oblique

This muscle lies deep to the external oblique muscle. Its fibres arise from the crest of the ilium and by a broad band of fascia from the spinous processes of the lumbar vertebrae. The fibres pass *upwards towards* the midline to be inserted into the lower ribs and, by an aponeurosis, to the linea alba. The fibres of this muscle are at right angles to those of the external oblique.

Figure 84

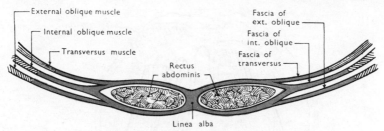

Diagram illustrating the arrangement of the fascia of the abdominal muscles.

Transversus abdominis

This is the deepest layer of muscle tissue of the abdominal wall. The fibres arise from the iliac crest and the lumbar vertebrae and pass across the abdominal wall to be inserted into the linea alba by an aponeurosis. The fibres of this muscle are at right angles to those of the rectus abdominis.

FUNCTIONS

The functions of the four muscles are to form a strong muscular anterior wall to the abdominal cavity. When the muscles contract together they have the effect of:

1. compressing the abdominal organs
2. flexing the vertebral column in the lumbar region.

Contraction of the muscles on one side only, bends the trunk towards that side. Contraction of the oblique muscles on one side, rotates the trunk.

THE INGUINAL CANAL

This is a canal 2·5 to 4 centimetres long which passes obliquely through the abdominal wall. It runs parallel to and is immediately in front of the inguinal ligament. In the male it contains the *spermatic cord* and in the female, the *round ligament*. This constitutes a weak point in the otherwise strong abdominal wall, but the fact that the canal runs obliquely through the wall does, to some extent, reduce the weakness.

Muscles of the Posterior Abdominal Wall

The muscles of the posterior part of the abdominal wall have been described above. They consist of:

    quadratus lumborum
    psoas
    internal oblique
    transversus abdominis.

Figure 85

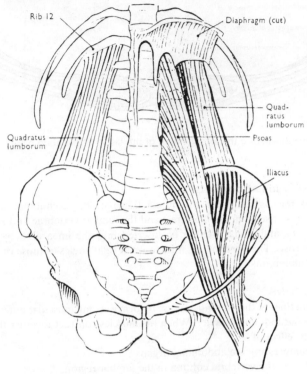

The muscles which form the posterior abdominal wall.

Figure 86

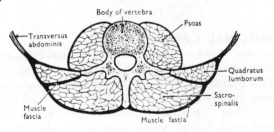

Diagram illustrating the muscles of the posterior abdominal wall in association with a lumbar vertebra.

Muscles of the Pelvic Floor

The pelvic floor is divided into two identical parts at the midline. It is formed by two pairs of muscles:

    levator ani   2

    coccygeus   2

Figure 87

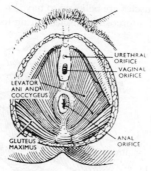

The muscles which form the pelvic floor.

Levator ani

This is a broad flat muscle which originates from the inner surface of the bones of the true pelvis. The two muscles unite at the midline and together they form a sling which supports the organs of the pelvic cavity.

Coccygeus

This muscle is triangular in shape and is situated behind the levator ani. It originates from the medial surface of the ischium and is inserted into the sacrum and coccyx.

These two muscles complete the formation of the pelvic floor, which is perforated in the male, by the urethra and the anus and in the female, by the urethra, vagina and anus.

Muscles of Respiration

The main muscles of respiration are:

external intercostal	11 pairs.
internal intercostal	11 pairs.
diaphragm	1

This group of muscles can be assisted during difficult or deep breathing by the muscles of the abdomen, neck and shoulders.

External intercostal muscles

These muscles occupy the spaces between the ribs. The fibres of each extend *downwards and forward* from the inferior border of one rib to the superior border of the rib below. Their action is to *raise the ribs upwards and outwards* thus increasing the size of the thoracic cage laterally and anteroposteriorly. Because of the increase in the capacity of the thoracic cage inspiration takes place.

Internal intercostal muscles

There are eleven pairs of these muscles which lie under the external intercostal muscles. Their fibres arise from the lower border of one rib and pass obliquely *downwards and backwards* to the upper border of the rib below. These muscles are believed to have an action antagonistic to that of the external intercostal muscles during forced expiration.

Figure 88

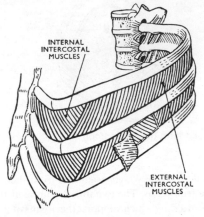

The intercostal muscles.

Diaphragm

The diaphragm is a dome-shaped musculo-fibrous structure which separates the abdominal and thoracic cavities, thus forming a concave roof for the abdominal cavity and a convex floor for the thoracic cavity.

Figure 89

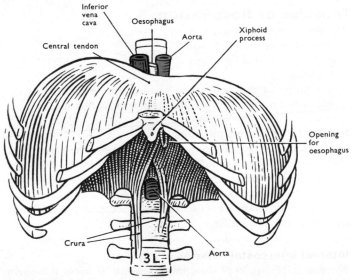

The diaphragm.

It is made up of *radiating muscle fibres* which are inserted at one end into a *central aponeurosis* and originate from the circumference of the trunk.

Anteriorly—by two slips from the ensiform process of the sternum.

Laterally—from the lower ribs.

Posteriorly—from two crura, or fibrous bands, from the first three lumbar vertebrae.

There are three main openings in the diaphragm for the passage of important structures between the thoracic and abdominal cavities.

Aortic Opening. This opening is actually posterior to the diaphragm, between the vertebral column and the diaphragm at the level of the 12th thoracic vertebra. Because of the importance of the aorta—it is the largest artery in the body—it is essential that it is not compressed in any way by the contraction of diaphragmatic muscle. The *thoracic duct,* the largest lymphatic vessel of the body, passes through the aortic opening.

Oesophageal Opening. This opening is situated anterior to the aortic opening at the level of the 10th thoracic vertebra. The oesophagus, or gullet, passes between the muscle fibres of the diaphragm and conveys food from the mouth to the stomach. Passing through this opening with the oesophagus are the vagus nerves which transmit nerve impulses to most of the abdominal organs.

Vena Caval Opening. This opening is in the central aponeurosis in front of the oesophageal opening at the level of the 8th thoracic vertebra. It allows for the passage of the inferior vena cava upwards to the heart. The inferior vena cava is one of the largest veins in the body. The phrenic nerves which supply the diaphragm pass through this opening.

Organs in relationship to the diaphragm

Superiorly: the right and left lungs with the heart between.

Inferiorly: the liver, the stomach, the spleen, two kidneys and two suprarenal glands.

Functions of the diaphragm

The diaphragm is the main muscle of respiration and its action is only partly under the control of the will.

When the radiating muscle fibres contract they *pull* on the central aponeurosis and *flatten* the dome-shaped structure, thus increasing the length of the thoracic cage and, as a result, inspiration occurs. When the muscle of the diaphragm is contracted it increases the pressure within the abdominal and pelvic cavities, and so assists in *micturition* (the passing of urine), *defaecation* (the passing of faeces), and *parturition* (childbirth). When the muscle relaxes it recovers its dome shape and expiration occurs.

The diaphragm and the external intercostal muscles contract *simultaneously* and so the size of the thoracic cage is increased laterally, antero-posteriorly and longitudinally at the same time.

5. The Blood

Blood is commonly described as a connective tissue. It provides one of the methods of communication between the cells of different parts of the body. For their normal functioning all cells require a supply of raw materials, such as, nutrients. As a result of all metabolic processes, using these raw materials, substances which are useful to the body and waste products which may be harmful, are produced. The blood is the medium of transportation of all of these substances.

Nutrient materials enter the blood by absorption from the alimentary tract, some waste products are conveyed to the kidneys for excretion from the body and some secretions influence the physiological activities of cells in other parts of the body

The blood constitutes approximately 8 per cent of the body weight or 5·6 litres in a 70 kg man.

Figure 90

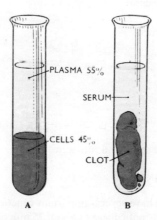

Blood in a test tube.

(A) Illustrating the percentage composition of plasma and cells.
(B) Illustrating the presence of a clot and serum.

Composition of Blood

The composition of blood must vary from time to time according to the activity and function of the organ it is supplying. The cells are continually subtracting from, and adding to, the constituents of blood according to their needs, but there are known normal limits within which these changes occur.

The blood is described as being composed of a faintly yellow transparent fluid known as the *plasma* and floating in this fluid are numerous *cells or corpuscles* of different kinds.

The plasma constitutes approximately 55 per cent, and the corpuscles approximately 45 per cent of the volume of blood. The specific gravity of the blood is about 1055.

THE PLASMA

The plasma is composed of the following constituents.
Water 90 to 92 per cent
Proteins in the plasma
    serum globulin
    serum albumin
    fibrinogen
    prothrombin.
Mineral salts (inorganic salts)
    sodium chloride and sodium bicarbonate. Also small amounts of potassium, magnesium, phosphorus, calcium, iron, copper and iodine.
Nutrient materials (from digested foods)
    amino acids
    glucose
    fatty acids and glycerol
    vitamins.
Organic waste products
    urea
    uric acid
    creatinine.
Hormones
Enzymes
Antibodies and antitoxins
Gases
    oxygen, carbon dioxide and nitrogen.

PLASMA PROTEINS

The plasma proteins have several important functions to perform.

1. They exert an osmotic force of about 25 mm of mercury across the capillary wall (*oncotic pressure*).

2. They give viscosity to the blood, thus assisting to some extent in the maintenance of the *blood pressure*.

Serum albumin is thought to be formed in the liver. *Serum globulin* is derived from the white blood cells known as the *lymphocytes*.

Fibrinogen is produced in the liver and is necessary for the clotting or coagulation of blood. (Plasma from which fibrinogen has been removed due to clotting is known as *serum*.)

Prothrombin is formed in the liver and is an essential substance in the mechanism of blood coagulation. (This mechanism is discussed on page 111).

MINERAL SALTS

These salts are necessary for the formation of cells, the contraction of muscles, the transmission of nerve impulses and to maintain the balance between acids and alkalis in the body. In health the blood is always *slightly alkaline* in reaction. This is expressed by the symbol pH, which represents the hydrogen ion concentration in the blood. The pH of the blood is maintained at about 7·4 by a complicated series of chemical changes occurring within the body.

Figure 91

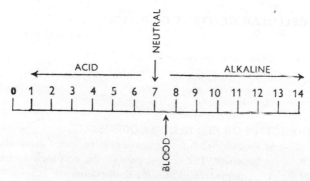

Diagrammatic illustration of the pH scale.

NUTRIENT MATERIALS

Amino acids, glucose, fatty acids, glycerol and vitamins are absorbed into the blood from the alimentary tract. These are the nutrient materials derived from the digestion of carbohydrates, proteins and fats and are required to maintain the functioning of the body cells.

ORGANIC WASTE PRODUCTS

Urea, uric acid and creatinine are waste products of protein metabolism. They are formed in the liver and are conveyed by the blood to the kidneys to be excreted.

HORMONES
Hormones are chemical substances which are formed by certain glands and are passed directly into the blood which transports them to other organs where they influence activity.

ENZYMES
Enzymes are chemical substances which can produce or speed up chemical changes in other substances without themselves being changed.

ANTIBODIES AND ANTITOXINS
These are protective substances consisting of complex proteins which are produced by *plasma cells* found mainly in lymph glands and in the spleen. Foreign proteins, such as micro-organisms or other large molecular substances, when introduced into the body may act as antigens and stimulate the production of specific antibodies. For example, antibodies which provide immunity against one antigen are not effective against another.

GASES
Small amounts of oxygen, carbon dioxide and nitrogen are dissolved in the plasma.

THE CELLULAR CONTENT OF BLOOD
There are three varieties of cells or corpuscles present in blood.

Erythrocytes or red blood corpuscles.

Leucocytes or white blood cells.

Thrombocytes or platelets.

ERYTHROCYTES OR RED BLOOD CORPUSCLES
These are described as circular bi-concave non-nucleated discs, about 7 microns* in diameter. The central part of the corpuscle is much thinner than the circumference, thus the term bi-concave.

The normal erythrocyte count is usually slightly lower in women than in men.

Women: 4·5 to 5 million per cubic millimetre of blood.

Men: 5 to 5·5 million per cubic millimetre of blood.

THE DEVELOPMENT AND LIFE HISTORY OF ERYTHROCYTES
The erythrocytes are formed in the *red bone marrow*. It will be remembered that red bone marrow is present in the cancellous bone at the extremities of long bones and between layers of compact bone in flat and irregular bones such as the sternum, skull, ribs and vertebrae.

* A micron (μ) is ·001 of a millimetre or about $\frac{1}{25000}$ of an inch.

Figure 92

Red blood corpuscles.

In addition to erythrocytes some leucocytes (granulocytes) and thrombocytes develop from the same parent cells, *the myeloblasts*.

The erythrocytes pass through several stages of development in the red bone marrow before they are mature and pass from the bone marrow into the circulating blood (see Fig. 93).

Formation

There are two main lines of development to the stage of mature erythrocyte. One is the maturation of the erythrocyte itself and the other the formation of a substance capable of transporting oxygen called *haemoglobin*. Erythrocytes entering the blood are not normal unless they develop satisfactorily along *both* lines.

The normal maturation of the cell itself requires the presence of a number of different chemical substances, one of the most important of which is *vitamin B_{12}* (the anti-anaemic factor). This substance is present in food and its absorption from the small intestine is dependent upon an unidentified substance secreted by the stomach called the *intrinsic factor*. When there is not sufficient vitamin B_{12} in the red bone marrow *pernicious anaemia* occurs.

Haemoglobin is a complex protein which is synthesised inside the immature erythrocyte in the red bone marrow. *Iron* is one of the chemicals necessary for its formation and if it is in short supply the individual develops *iron-deficiency anaemia*. The normal haemoglobin levels in the blood are slightly different in men and women:

Men	16 g per 100 ml blood.
Women	14 g per 100 ml blood.

Amounts of iron and vitamin B_{12} absorbed from the alimentary tract which are in excess of immediate requirements are stored in the liver.

Erythropoiesis, or the development of red blood cells, appears to be controlled by what is known as a feed-back mechanism. This means that the body produces erythrocytes at the same rate as they are destroyed.

Figure 93

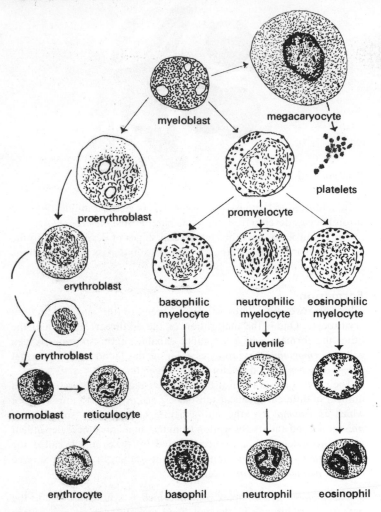

Development of blood cells in the red bone marrow.

If there is an insufficient supply of oxygen to the body cells, for example, following haemorrhage or because an individual lives at a high altitude where the oxygen pressure in the atmospheric air is reduced, the bone marrow is stimulated to increase its production of erythrocytes. Deficiency of oxygen (hypoxia) as such, and a hormone called *erythropoietin,* probably produced by the kidneys, are the two main stimulants. After haemorrhage it is normal to find that there are a number of immature cells, reticulocytes, in the blood.

Destruction

The life span of erythrocytes is believed to be about 120 days. The breakdown of these cells is called *haemolysis* and it occurs mainly in the spleen. During haemolysis iron is extracted and reused for haemoglobin synthesis in the red bone marrow. The protein part of the erythrocyte is converted to a substance called *biliverdin* which is almost completely changed to *bilirubin*, a yellow pigment. While circulating in the blood bilirubin is bound to the plasma protein albumin. In the liver it is again changed to what is described as a *conjugated form* and excreted in bile as bile pigment. When tests are carried out to find the level of bilirubin in the blood the *unconjugated bilirubin* is described as *indirect-reacting* and the *conjugated bilirubin* as *direct-reacting*. When there is excess bilirubin in the blood the skin, the whites of the eyes and the mucous membrane become yellow. This is known as *jaundice* and the cause may be identified by measuring the direct- and indirect-reacting bilirubin.

Functions

The haemoglobin in the erythrocytes is concerned mainly with the transportation of oxygen from the lungs to all the cells of the body.

LEUCOCYTES OR WHITE BLOOD CELLS

The leucocytes differ from the erythrocytes in that they possess a nucleus, are fewer in number and larger in size. They vary in size from about 8 to 15 microns and there are 6,000 to 10,000 per cubic millimetre of blood.

The leucocytes are divided into two main varieties:

granular or polymorphonuclear leucocytes
nongranular or mononuclear leucocytes.

Granular leucocytes

These cells develop in the red bone marrow and pass through several stages of development before entering the blood stream. They originate from the same parent cells as the erythrocytes, i.e., myeloblasts (see Fig. 93). As the cells develop they develop granules in the cytoplasm and they retain their nucleus.

The granulocytes constitute about 75 per cent of all the white blood cells and three varieties are described.

Neutrophils   approximately 70 per cent
Eosinophils   approximately 4 per cent
Basophils   approximately 1 per cent.

These names are derived from the fact that the *granules* of the different groups when stained in the laboratory either absorb an acid dye such as *eosin* which is red, or an alkaline or basic dye *methylene blue*, or absorb both acid and alkaline dyes thus producing a neutral purple colour.

Figure 94

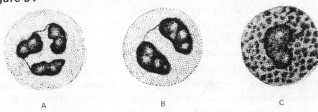

Diagram showing the three different types of granulocytes.

A. Neutrophils
B. Eosinophils
C. Basophils.

Functions of granular leucocytes

Neutrophils. The neutrophils are responsible for providing the body with a defence against invading micro-organisms. They are attracted in large numbers to any area of the body which has been invaded by micro-organisms. They are attracted by chemical substances liberated by the infected tissue (chemotaxis).

Figure 95

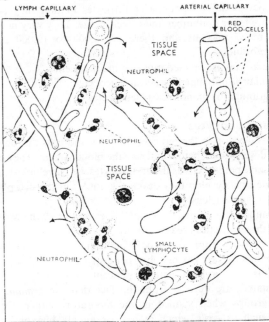

Scheme showing neutrophils leaving blood capillaries by the process of diapedesis.

Neutrophils leave the blood by insinuating themselves through the walls of the capillaries in the infected area (a process known as *diapedesis*). Thereafter they ingest and kill the organisms by digesting them (process known as *phagocytosis*). The pus which may exude from an infected area consists of destroyed tissue, live micro-organisms and dead neutrophils which have ingested more micro-organisms than they could digest.

The life span of neutrophils which remain in the blood vessels is believed to be about 30 hours and those which leave the blood do not return.

Eosinophils and Basophils. Their functions are not completely understood. They are less actively motile than the basophils. It is thought that the eosinophils phagocytose the particles which are formed when antigens and antibodies react. In allergic conditions, such as asthma, they are increased in numbers.

The numbers of eosinophils and basophils are reduced when there is an increase in the amount of the hormone hydrocortisone in the blood.

Little is known of the functions of basophils.

Figure 96

MICRO-ORGANISMS

MICRO-ORGANISMS INGESTED BY NEUTROPHIL

Scheme showing phagocytic action of neutrophils.

Nongranular leucocytes

There are two varieties of nongranular leucocytes and they comprise about 25 per cent of all leucocytes.

Lymphocytes, large and small.

Monocytes.

Figure 97

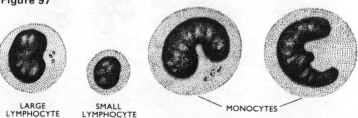

LARGE LYMPHOCYTE   SMALL LYMPHOCYTE   MONOCYTES

Large and small lymphocytes.

Lymphocytes. The large and small lymphocytes closely resemble each other and differ only in size. These cells consist of a large single nucleus and very little cytoplasm. Approximately 92 per cent of all lymphocytes are small.

The lymphocytes are produced in the lymphatic tissue of the body, that is, in the lymphatic glands and in the lymphoid tissue which is present in the spleen, liver and many other organs. Lymphocytes are known to cross the walls of the blood capillaries into tissue fluid. How this process occurs is not fully understood but it is known to be different from the diapedesis of neutrophils.

The Function of Lymphocytes. This is still not completely understood, but it is thought that they are concerned with the process of immunity.

They are associated with the production of serum globulin, one of the plasma proteins which is thought to produce antibodies and antitoxins, thus preventing specific infections. Lymphocytes play an important part in the immunological reactions to tissue transplantation.

Monocytes. These are large cells with a large nucleus, they are even bigger than the large lymphocytes and are relatively few in number. They are thought to originate from a system of primitive cells, the *reticulo-endothelial system.* This tissue is found in organs such as the liver, spleen, lungs and lymphatic glands.

Figure 98

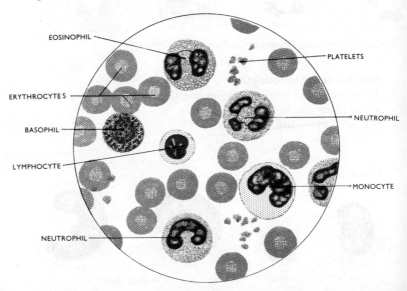

Normal blood cells.

Function of the Monocytes. Their function closely resembles that of the neutrophils in that they are actively motile, phagocytic in action and will leave the blood stream to ingest micro-organisms and other foreign material which may be introduced into the tissues.

THROMBOCYTES OR PLATELETS

These are extremely minute cells very much smaller than the erythrocytes. They do not possess a nucleus, but have granules in their protoplasm. They have their origin in the red bone marrow and are derived from myeloblasts (see Fig. 93).

There are approximately 300,000 thrombocytes in a *cubic millimetre of blood.*

The function of thrombocytes

They play a very important part in the clotting of blood.

The Clotting or Coagulation of Blood

When any blood vessel is damaged blood will escape. To prevent blood loss the body reacts by a mechanism which is termed the *coagulation or clotting of blood.* This is a very complex process and is described here in its simplest terms.

There are some substances which must be present before blood clotting occurs:

    prothrombin
    calcium
    fibrinogen
    thromboplastin.

Prothrombin, calcium and fibrinogen are all normal constituents of the blood. *Thromboplastin*, on the other hand, is released only when a blood vessel and tissue cells are damaged, for example, by a cut, thromboplastin is liberated from the injured tissue cells and from the damaged thrombocytes. The release of thromboplastin starts off a series of events which will eventually produce a jelly-like mass known as a blood clot.

Prothrombin as such is inactive but when acted upon by *thromboplastin* in the presence of *calcium* it is converted into an active substance known as *thrombin.*

Thrombin in turn acts upon *fibrinogen* which is converted into an insoluble thread-like structure known as *fibrin.*

The fibrin threads enmesh the cellular content of the blood, that is, the erythrocytes and the leucocytes. This combination of fibrin and blood cells occludes the opening in the wall of the blood vessel.

After a time the clot shrinks and a clear sticky fluid is released. This is *serum*, i.e., plasma from which fibrinogen has been removed.

The mechanism of clotting can be expressed simply by the following formula:

prothrombin + calcium + thromboplastin = thrombin
   (inactive)   (from damaged   (active)
                                 tissue cells and
                                 platelets)

thrombin + fibrinogen = fibrin (fine threads).
fibrin + blood cells = clot.

FACTORS WHICH AFFECT THE CLOTTING OF BLOOD

Vitamin K

Vitamin K is necessary for the satisfactory formation of prothrombin which is made in the liver. Because of this, vitamin K is sometimes termed the *anti-haemorrhagic vitamin*. It is obtained from green foods such as cabbage, spinach and cauliflower. Some vitamin K is formed by bacterial action in the large intestine but it is doubtful if any of this is absorbed.

Heparin

This is a substance which prevents the coagulation of blood and is therefore known as an *anti-coagulant*. Heparin is found in circulating basophils and in *mast cells*. These are wandering cells which are found in most tissues and in profusion in loose connective tissue. There is a substantial amount of heparin secreted, by mast cells and basophils, into loose connective tissue *surrounding* the capillaries which prevents coagulation. Heparin prevents the conversion of prothrombin to thrombin. It is used clinically as an anti-coagulant.

Blood Groups

Blood transfusions are frequently carried out in hospital wards. However, blood may not be taken indiscriminately from one person and transfused into another.

The membranes of erythrocytes contain antigens called *agglutinogens* and some people have natural antibodies, *agglutinins* in their plasma which react against specific agglutinogens.

Individuals are divided into four main blood groups, A, B, AB and O. Group A blood always contains agglutinins which cause clumping or *agglutination* of Group B blood. Similarly Group B blood causes agglutination of Group A blood. Individuals with Group AB blood have no agglutinins and those with Group O blood have agglutinins against both Group A and Group B blood. Figure 99 summarises the compatibility of the different blood groups.

If incompatible blood is transfused the clumps of agglutinated erythrocytes block the small blood vessels and eventually they are haemolysed.

Figure 99

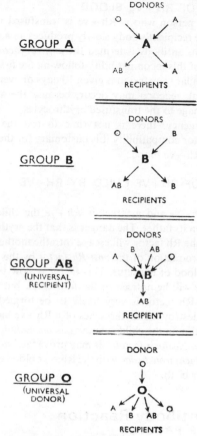

GROUP **A**

DONORS

RECIPIENTS

GROUP **B**

DONORS

RECIPIENTS

GROUP **AB**
(UNIVERSAL RECIPIENT)

DONORS

RECIPIENT

GROUP **O**
(UNIVERSAL DONOR)

DONOR

RECIPIENTS

Blood groups showing their compatibility.

The frequency of occurrence in Caucasians, of the *four blood groups* is as follows:

group A	approximately 42 per cent	
group B	approximately 8 per cent	of the population.
group AB	approximately 4 per cent	
group O	approximately 46 per cent	

The Rhesus Factor

Since the above blood groups were described further agglutinogens have been discovered. One of the most important of these is the Rhesus (Rh) factor, named after the rhesus monkey because it was used in the original investigations.

In about 85 per cent of Caucasians the Rh factor is present (Rh positive) and in the remaining 15 per cent it is absent (Rh negative). In Africans, Chinese and Japanese 99 to 100 per cent are Rh positive.

TRANSFUSION OF RH+VE BLOOD

If the blood of a person who is Rh+ve is transfused into a person who is Rh−ve the recipient's body slowly produces an antibody to the agglutinogens. This antibody is termed the *anti-Rh factor*. There may be no indication of this incompatibility following the first transfusion, but if a second similar transfusion is given 10 days or even years later a serious, often fatal, reaction may occur, because the anti-Rh factor causes severe damage to the transfused erythrocytes.

If, in an emergency, there is not time to test the donor's and recipient's blood for compatibility it is customary for the patient to be given Group O, Rh−ve blood.

CONCEPTION OF RH+VE CHILD BY RH−VE MOTHER

If a mother is *Rh−ve* and a father is *Rh+ve* the child may inherit the Rh factor from its father. The danger is that the erythrocytes of the fetus containing the Rh factor will escape into the mother's circulation stimulating the production of the *anti-Rh factor* by the mother which passes into the blood of the fetus. This factor is slow to develop and no serious effects will be noticed in the first child, but in subsequent pregnancies anti-Rh factor is very likely to be formed in quantities large enough to destroy the erythrocytes of a Rh+ve baby. When this occurs the condition in the child is known as *haemolytic disease of the newborn* or *erythroblastosis foetalis*. It may prove fatal unless the infant is given a *replacement transfusion* with the Rh−ve blood either before or immediately after birth.

Antigen Antibody Reaction
THE IMMUNE PROCESS

The immune response is the effect on the body of contact with antigens or substances of various types which are 'foreign' to the individual. Antibodies which are specific to the individual antigens, and will neutralise them, are produced by the lymphocytes and lymphoid tissue in the lymph glands and in the spleen. Some of the antigens which stimulate such reactions in the body are:

micro-organisms
protein molecules from animals
pollen from plants and flowers
polysaccharides
foreign tissues, for example, transplanted organs
drugs, for example, large molecule drugs such as penicillin.

After the first contact with an antigen there is an interval of about two weeks before antibodies are found in the blood. During this period there is intense activity within the lymphoid tissue. The antibody concentration in the blood which results from this *primary response* does not reach a high level and does not persist unless a second dose

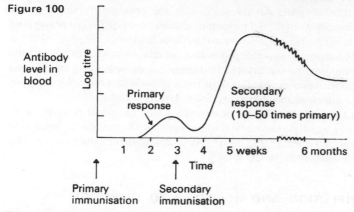

Figure 100

The antibody response.

of antigen is encountered. As a result of this *secondary response* there is a marked and rapid increase in the antibody level in the blood which is sustained for a considerable period of time (see Fig. 100).

Once the individual has responded to an antigen the cells in the lymphoid tissue retain a 'memory' of the antigen and even after several years a further exposure will result in a secondary response.

Individuals are always exposed to antigens in the air, for example, pollen from plants. In this way over a period of years the human body builds up a variety of antibodies to many naturally occurring antigens.

ALLERGIC REACTIONS OR HYPERSENSITIVITY

An allergic reaction is an undesirable response by the body to the invasion of a particular antigen which has previously been encountered.

The initial number of antibodies formed in response to an antigen may be small. Some of these bodies may remain in the circulation and some may be absorbed by the mast cells.

If an individual is exposed to the same antigen again there may be sufficient antibodies in the circulation to neutralise the antigen, but if large amounts of the antigen are inhaled, eaten or come in contact with the skin there may not be sufficient *circulating* antibodies to neutralise them. The antigens may then combine with the antibodies *within the cells*. When this antigen-antibody reaction takes place within the cells a so called *allergic reaction* occurs.

These reactions are characterised by a multiplicity of clinical features such as:

headaches
skin rashes
oedema of the arms and legs
angio-neurotic oedema (swelling of the lips and face)
vomiting and diarrhoea
rhinitis, sinusitis and catarrh.

The antigen-antibody reaction in the cells greatly increases the permeability of the cell membrane and may damage the cell protoplasm releasing *histamine* into the extracellular fluid.

Circulating histamine promotes vaso-dilatation, venous pooling and diminished venous return. Histamine causes oedema of the mucosa and increases the tone of most types of smooth muscle, for example, the muscle of the alimentary and respiratory tracts.

The most serious allergic reaction which can occur is *anaphalactic shock* which may prove fatal within minutes. Death is due to insufficient venous return to the heart. As a result the cardiac output falls immediately to 'shock' level.

INFECTION AND REACTION TO INFECTION

Infection

Infection occurs when micro-organisms which invade the body, grow, reproduce, multiply and produce toxins.

The human body is continuously exposed to contact with a wide variety of micro-organisms, such as, bacteria, fungi, protozoa and viruses. The micro-organisms which cause disease in man are described as *pathogenic*. Not all micro-organisms are pathogenic. There are large numbers which are not only *non-pathogenic* but are beneficial to man. *Commensal* organisms are non-pathogenic in their normal habitat but become pathogenic if they gain access to another part of the body.

The blood and tissue fluids of the body contain sufficient nutrient material for the survival, growth and reproduction of pathogenic micro-organisms. However, the normal healthy body possesses a variety of defence mechanisms which enable it to withstand the invasion and multiplication of micro-organisms.

DEFENCE MECHANISMS OF THE BODY

The individual is in contact with many pathogenic micro-organisms which may not gain entry to the body or which may be killed soon after entry and before they can multiply and cause disease. There is a number of non-specific and specific protections against invading organisms.

NON-SPECIFIC DEFENCE MECHANISMS

Skin and mucous membrane

When skin and mucous membrane are healthy and intact they provide a physical barrier to invading micro-organisms. The outer horny layer of the skin can be penetrated by only a few micro-organisms and the mucus secreted by the mucous membrane traps organisms on its sticky surface. Sebum and sweat, which are secreted on to the surface of the skin, contain bactericidal and fungicidal substances.

Bactericidal substances in body secretions

Hydrochloric Acid. This is present in high concentrations in gastric juice (pH1–2), and kills the majority of ingested organisms.

Lysozyme. This is a small molecule protein with bactericidal properties which is present in tears, in granulocytes and in most body secretions except sweat, urine and cerebro-spinal fluid.

Saliva. Saliva is secreted into the mouth and washes away food debris which would serve as culture media for micro-organisms. Its slightly acid reaction (pH about 6·7) inhibits the growth of some organisms.

Phagocytosis

When micro-organisms invade the body they immediately stimulate activity in neutrophils, monocytes and large cells described as macrophages which are found in the reticulo-endothelial cells of the liver, spleen, bone marrow and in connective tissue. The processes of phagocytosis were described on page 109.

SPECIFIC DEFENCE MECHANISMS

Acquired immunity

Acquired immunity depends upon the numbers of *antibodies* which are *specifically* produced to destroy *specific* micro-organisms and their toxins.

Antibodies appear to belong to the globulin group of proteins and are formed by lymphocytes and cells which are present in the lymph nodes and the reticulo-endothelial cells in the spleen and liver.

Antibodies are formed specifically, that is, they are produced in response to the invasion of *one specific type of micro-organism* which acts as an antigen. If the virus of measles invades the body antibodies are formed to destroy this particular pathogen. If the diphtheria bacillus invades the body another type of antibody is produced to destroy this particular pathogen. In this way the body produces a multiplicity of antibodies which provide a resistance to infection.

IMMUNITY

Immunity is said to occur when there are sufficient antibodies within the body to prevent the successful invasion of a particular type of pathogenic micro-organism.

Immunity can be acquired *naturally* or *artificially*.

In turn both these forms of immunity may be acquired *actively or passively*. When immunity is acquired actively the individual has responded to the antigen and produces his own antibodies. Passive immunity occurs when the individual has been given specific antibodies which have been produced by someone else.

Active naturally acquired immunity

This occurs when the body is involved *actively* in producing antibodies. This type of immunity can be acquired in two ways.

1. *By Actually Having the Disease.* The micro-organisms invade the body successfully where they grow and reproduce in sufficient numbers to produce the clinical features of a disease. During the course of the disease antibodies are produced in sufficient numbers to overcome the micro-organisms and the person recovers. These antibodies remain as a future protection against this particular pathogen.

2. *By Having Subclinical or Subliminal Infections.* In this instance the body is exposed to minute numbers of micro-organisms which are insufficient to give rise to recognisable disease but are sufficient to stimulate the production of antibodies.

Passive naturally acquired immunity

In this instance the antibodies are passed from the mother to the fetus through the placenta before the baby is born. This type of immunity is thought to be very short lived.

Active artificially acquired immunity

This type of immunity develops in response to the injection into the body of a suspension of killed or attenuated micro-organisms or by the injection of detoxicated toxins. *Attenuation* means that the organisms retain their antigenic identity, but can not cause the disease. This is achieved by heat treatment, exposure to chemicals such as formalin or by growing successive generations over a long period of time on artificial media in the laboratory.

The suspensions of micro-organisms which are dead or attenuated are called *vaccines* and the detoxicated toxins are called *toxoids.*

Vaccines and toxoids are prepared so that they will not cause the disease but are sufficiently powerful to stimulate the production of antibodies and thus build up an active immunity.

Many diseases can be prevented by artificial active immunisation, some of these are:

tuberculosis	poliomyelitis
yellow fever	typhoid fever
tetanus	whooping cough
measles	diphtheria
smallpox	anthrax.

Passive artificially acquired immunity

In passive immunity the individual plays no active part in the production of antibodies.

The ready-made antibodies obtained from human or animal serum are injected into the individual. Human serum contains the *ready-made antibodies* if the individual has recovered from an infectious disease, for example, measles. Animal serum contains the antibodies following the active immunisation of an animal.

Animal anti-sera may be used both *prophylactically* (to prevent infection) or *therapeutically* (to treat an infection).

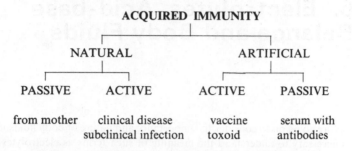

ACQUIRED IMMUNITY

NATURAL   ARTIFICIAL

PASSIVE   ACTIVE   ACTIVE   PASSIVE

from mother   clinical disease   vaccine   serum with
          subclinical infection   toxoid   antibodies

Summary of the Functions of the Blood

1. The blood transports oxygen as oxyhaemoglobin from the lungs to the body cells, and returns carbon dioxide from the cells to the lungs for excretion.

2. The blood is the means whereby all nourishment is transported to the cells. This nutritive material is absorbed into the blood stream from the small intestine in the form of glucose, amino acids, fatty acids, glycerol, vitamins, mineral salts and water.

3. The blood removes all waste products from the tissues and cells. These waste materials are transported to the appropriate organs for excretion or to the liver to be prepared for excretion. After preparation by the liver the waste products of protein metabolism are in the form of urea, uric acid and creatinine and as such, they are transported in the blood to the kidneys for excretion.

4. The blood transports hormones and enzymes to the cells.

5. The blood aids in the defence of the body against the invasion of micro-organisms and their toxins due to:

     (a) the phagocytic action of neutrophils and monocytes

     (b) the presence of antibodies and antitoxins.

6. By the mechanism of clotting, loss of body fluid and loss of blood cells is prevented.

7. The blood helps to maintain the body temperature. Due to the activity of the cells and tissues heat is produced and the circulating blood is warmed. If too much heat is produced the blood vessels near the surface of the body dilate and heat is lost by radiation, conduction, convection and evaporation of sweat. If the temperature of the outside atmosphere is low the superficial blood vessels constrict and heat loss is prevented.

6. Electrolytes, Acid-base Balance and Body Fluids

Before going on to consider the importance and nature of body fluids it is necessary to understand the meaning of such terms as electrolytes, isotopes, pH and buffer substances. If these terms are to be understood it is essential to survey some of the principles of the physics and chemistry of atoms, elements and compounds. Broadly speaking, the chemical compounds dissolved in the water of the body can be divided into organic and inorganic compounds.

Before discussing the compounds found in solution in the water of the body, the atom, and its main constituent parts, has to be considered.

THE ATOM
The atom is described as the smallest particle of an element which can take part in a chemical change and all atoms of one element are identical. However, the atoms of any one element are different from those of all other elements. This will be more clearly understood when the structure of the atom has been described.

The structure of the atom
A considerable number of particles have been described as constituting the structure of the atom, three of which will be discussed here.

1. *Protons.* Protons are particles which exist in the nucleus, or central part of the atom. Each proton is described as having one unit of positive electrical charge and one unit of mass.

2. *Neutrons.* Neutrons are also found in the nucleus of the atom. They have no electrical charge and one unit of mass.

3. *Electrons.* Electrons are particles which revolve in orbit around the nucleus of the atom at a distance from it, as the planets revolve round the sun. Each electron carries one unit of negative electrical charge and its mass is so small that it can be disregarded when considering the weight of an atom as a whole.

In all atoms the number of positively charged protons in the nucleus is *equal* to the number of negatively charged electrons in orbit around the nucleus.

The difference which exists between elements is to be found in the *numbers* of these essential particles which make up their atoms. The planetary electrons revolve in concentric rings or shells around the nucleus and there is an optimum number of electrons in each shell.

Figure 101

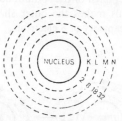

Schematic diagram of the nucleus and the first four electron shells of an atom. The shells are given the symbols K L M N.

Atomic number

The atomic number of an element is equal to the number of planetary electrons in orbit around the nucleus of the atoms of that element, for example, the atomic number of oxygen is 8 because it has 8 planetary electrons.

Figure 102

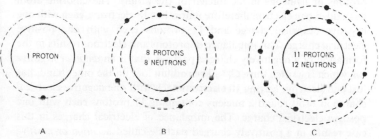

Schematic diagram of the structure of the atoms showing nuclear protons and neutrons and orbiting electrons.

(A) Hydrogen   (B) Oxygen   (C) Sodium.

Formation of compounds

It was mentioned earlier that the atoms of each element have a specific number of electrons in orbit around the nucleus. When the number of electrons in the outer shell of an element is the optimum number, that element is described as inert or chemically unreactive and will not easily form compounds by combining with other elements. The inert gases, neon, argon, krypton, xenon and radon come into this category.

Elements which have incomplete outer shells of electrons are reactive and will combine with other elements which also have incomplete outer electron shells. In the formation of *electrovalent or ionic compounds* electrons are transferred from one element to another. For example, when sodium (Na) combines with chlorine (Cl) to form sodium chloride (NaCl) there is the transfer of one electron, the only electron, from the outer shell of the sodium atom to the outer shell of the chlorine atom. This makes the outer shell (the M shell) of the chlorine up to its full capacity of 8 electrons.

Figure 103

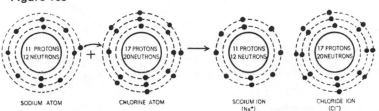

SODIUM ATOM   CHLORINE ATOM   SODIUM ION (Na⁺)   CHLORIDE ION (Cl⁻)

Schematic diagram of the formation of the ionic compound sodium chloride.

The number of electrons is the only change which occurs in the atoms in this type of reaction. There is no change in the number of protons or neutrons in the nucleii of the atoms. The chlorine atom has gained one electron therefore it now has 18 electrons, each with one negative electrical charge and 17 protons each with one positive electrical charge. This imbalance of protons and electrons results in the formation of a negatively charged particle, that is an *anion or negative ion*, which is written thus, Cl^-. The sodium ion, on the other hand, has lost one electron, leaving 10 electrons each with one negative electrical charge in orbit round a nucleus containing 11 protons each with one positive electrical charge. The imbalance of electrical charges in this case results in a positively charged particle called a *cation or positive ion*, which is written Na^+. The number of electrons which transfer in the formation of ionic compounds is indicated by the number of superscript plus or minus signs, for example, the magnesium ion is written Mg^{++} and the sulphide ion S^{--}.

The nature of electrical charge which is used daily in lighting, heating and so on, is closely related to the structure of the atom. Power initiated by water falling from a height or by steam, produced when water is heated within a confined space, is used to drive a dynamo which generates electricity, which is the movement of electrons. The electric current is then led away along a metal wire called a *conductor*. A substance which prevents the flow of electricity, or electrons, is a *non-conductor* or *insulator*.

Faraday, in *c.* 1830, discovered that some substances when in solution conducted electricity while others did not. It was later discovered that the solutions which conducted electricity were ionic compounds. The solutions which conduct electricity are called *electrolytes* and those which do not are called *non-electrolytes*.

When sodium chloride is dissolved in water it is said to ionise or dissociate into a number of anions (Cl^-) and an equal number of cations (Na^+). Sodium chloride is present in the human body in solution in water and is therefore an electrolyte. Glucose, on the other hand, does not dissociate when dissolved in water and is therefore not an electrolyte.

In this discussion, sodium chloride has been used as the example of the formation of an ionic compound and to illustrate electrolyte activity. There are, however, many other electrolytes within the human body which, though in relatively small quantities, are equally important. Although these substances may enter the body in the form of compounds, such as sodium bicarbonate, they are usually discussed in the ionic form, that is, the sodium ion is written Na^+ and the bicarbonate ion HCO_3^-.

The bicarbonate part of sodium bicarbonate is derived from carbonic acid (H_2CO_3). All inorganic acids contain hydrogen which is combined with another element or with a group of elements which act like a single element. These groups are called *radicals*. Hydrogen combines with chlorine to form hydrochloric acid (HCl) and with the phosphate radical to form phosphoric acid (H_3PO_4). When these two acids ionise they do so thus:

$$HCl \longrightarrow H^+ Cl^-.$$
$$H_3PO_4 \longrightarrow 3H^+ PO_4^{---}.$$

In the second example three atoms of hydrogen have each lost one electron and the three electrons have been taken up by one unit, the phosphate radical, to make a phosphate ion with three negative charges.

Isotopes

In an over simplification it was stated earlier that all atoms of an element are identical. This was true in so far as the number of protons and electrons are concerned, but some atoms have a different number of neutrons in the nucleus. This does not affect the electrical activity of these atoms but it does affect their weight. For example, there are three forms of the hydrogen atom. The most common form has one nuclear proton and one orbiting electron, while a second form has one orbiting electron, and one proton and one neutron in the nucleus, and occurs in 1 in 5,000 atoms of hydrogen. A third form is more rare, 1 in 1,000,000 atoms. It has one orbiting electron, and in its nucleus one proton and two neutrons. The three forms of hydrogen are called *isotopes* of hydrogen.

Figure 104

 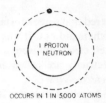 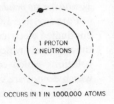

MOST COMMON FORM   OCCURS IN 1 IN 5,000 ATOMS   OCCURS IN 1 IN 1,000,000 ATOMS

Schematic diagram of the isotopes of the hydrogen atom.

Atomic weight

The atomic weight of an element is the sum of the number of protons and the number of neutrons in the nucleus of the atoms of the element. Taking into account the isotopes of hydrogen and the proportions in which they occur, the atomic weight of hydrogen is 1·008, although for most practical purposes it can be taken as 1.

Oxygen has an atomic weight of 16. This means that a specific number of atoms of oxygen weigh 16 times more than the same number of atoms of hydrogen. Any standard scale of weight may be used for this purpose, for example, milligrammes, grammes, pounds or ounces.

Chlorine has an atomic weight of 35·5, this is because it exists in two forms, one form having 18 neutrons in the nucleus, and the other 20 neutrons in the nucleus. Because the proportions of these two forms are not equal, the *average atomic weight* which emerges is 35·5.

Molecular weight

The molecular weight of a compound is the sum of the atomic weights of the elements which form the molecules of the compound. For example:

a. water (H.OH)	2 hydrogen atoms	2
	1 oxygen atom	16
	molecular weight =	18
b. sodium bicarbonate (NaHCO₃)	1 sodium atom	23
	1 hydrogen atom	1
	1 carbon atom	12
	3 oxygen atoms	48
	molecular weight =	84

Equivalent weight

If someone in a laboratory wished to synthesise water he would find that, to prevent waste, he would require to take twice as much hydrogen as oxygen, and provide the appropriate conditions for chemical combination, to form water. Thus two volumes of hydrogen are *equivalent to* one volume of oxygen.

If hydrochloric acid (HCl) is to be made, it would be necessary to take 1 gramme of hydrogen and 35·5 grammes of chlorine. One gramme of hydrogen contains the same number of atoms as 35·5 grammes of chlorine. Provided these two gases were given the appropriate conditions for chemical combination, hydrochloric acid would be formed, and no residual hydrogen or chlorine left. It can be said, therefore, that 1 atomic weight of hydrogen is *equivalent to* 35·5 atomic weights of chlorine.

The definition of equivalent weight is the weight of an element which will combine with *one atomic weight of chlorine* or displace *one atomic weight of hydrogen*.

Using hydrochloric acid as an example, it can be seen that the equivalent weight of both elements is one, but if hydrochloric acid is combined with calcium it is found that one atomic weight of calcium combines with two atomic weights of chlorine releasing two atomic weights of hydrogen. The equation for this reaction is:

$$Ca + 2HCl \rightarrow CaCl_2 + H_2$$

The weight of calcium, therefore, which is equivalent to one atomic weight of chlorine is half an atomic weight of calcium (calcium: atomic weight = 40·8, therefore its equivalent weight = 20·4).

Any scale of measurement of the weight of these elements can be used provided the same scale is used for all the elements involved in the chemical reaction being considered. The measures commonly used are grammes or milligrammes or, for even smaller quantities, microgrammes.

The electrolytes which are found within the human body are usually expressed in terms of *milliequivalents per litre of body fluids (mEq/l)*. This is the equivalent weight in milligrammes in each litre of body fluid. Alternatively the body electrolytes may be expressed in *milligrammes per 100 millilitres of body fluid (mg%)*. The former scale of measurement is more useful because it provides figures for different electrolytes which are directly comparable. It is a measure of the number of ions, or chemically active units, present.

Examples of normal electrolyte levels in the body:

chlorine	97–106 mEq/l	or	560–620 mg%
sodium	135–147 mEq/l	or	310–340 mg%
potassium	3·5– 5·5 mEq/l	or	14– 22 mg%
calcium	4·5– 5·5 mEq/l	or	9– 11 mg%
phosphate	1·5– 2·0 mEq/l	or	2·5– 5 mg%
CO_2 combining power	23– 34 mEq/l	or	55– 75 mg%

pH or Hydrogen Ion Concentration

The number of hydrogen ions present in a solution is a measure of the acidity of the solution. The maintenance of the normal hydrogen ion concentration within the body is an important factor in the environment of the cells.

A standard scale for the measurement of the hydrogen ion concentration in solution has been developed. All acids do not ionise completely when dissolved in water, that is, all the molecules of acid in solution do not ionise and exist in the solution as electrically charged particles. The hydrogen ion concentration is a measure, therefore, of the amount of *dissociated acid* rather than of the amount of acid present. Strong acids dissociate more freely than weak acids, for example, hydrochloric acid dissociates freely into H^+ and Cl^-, while carbonic acid dissociates much less freely into H^+ and CO_3^-. The number of *free hydrogen ions* in a solution is a *measure of its acidity* rather than an indication of the type of molecule from which the hydrogen ions originated.

The alkalinity of a solution is dependent upon the number of hydroxyl ions (OH^-) present. Water is a neutral solution because every molecule contains one hydrogen ion and one hydroxyl radical. For every molecule of water (H.OH) which dissociates, one hydrogen ion (H^+) and one hydroxyl ion (OH^-) is formed, each one neutralising the other.

The scale for measurement of pH was developed taking water as the standard. It was found by experiment that 1 molecule in 550,000,000 molecules of water ionises into a hydrogen ion and a hydroxyl ion. This is the same proportion as 1 gramme hydrogen ion in 10,000,000 litres of water. Therefore 1 litre of water contains $\frac{1}{10,000,000}$ of a gramme of hydrogen ion. In order to make these figures more manageable this can be written:

1 litre of water contains $\frac{1}{10^7}$ g hydrogen ion.

$(10 = 10^1; 100 = 10^2; 10,000 = 10^4)$.

The fraction $\frac{1}{10^7}$ may be written 10^{-7}, the negative power indicates that this is a fraction. $\frac{1}{10,000,000}$ may be written $\frac{1}{10^7}$ or 10^{-7}.

Later, for every-day use, only the 'power' figure was used and the symbol pH placed before it.

Thus a neutral solution such as water, where the number of hydrogen ions is balanced by the same number of hydroxyl ions, the pH = 7. The range of this scale is from 0 to 14. If the pH is 0 it would mean that 1 litre of water contained $\frac{1}{1} = 1$ gramme of hydrogen ion; or at the other end of the scale if there were no hydrogen ions present it would be written $\frac{1}{10^{14}}$ or pH 14. It will be noted that a change of pH of one at any level in the scale means an increase or decrease by a factor of 10 in hydrogen ion concentration.

pH reading below 7 indicate an *acid solution*, while readings above 7 indicate an *alkaline solution*.

Ordinary litmus paper indicates that a solution is acid or alkaline by blue colouring for alkaline and red colouring for acid. It is possible to obtain specially treated absorbent paper which gives an approximate measure of pH by a colour change. Laboratories which have to make accurate measurements use a *pH meter*.

Figure 105

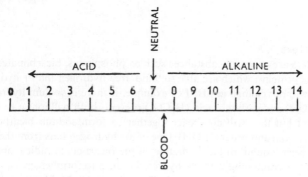

Diagrammatic illustration of the pH scale.

pH values of the body fluids

All body fluids have pH values which must be maintained within relatively narrow limits and within which the cells function normally. These pH values are not the same in every part of the body, for example, the pH values of the following secretions are:

blood	7·4 to 7·5
bile	7·8 to 8·6
pancreatic juice	8·0
saliva	6·4 to 6·8
urine	5·5 to 7·0
gastric juice	1·5 to 1·8

It can be seen that there is a wide variation of pH values within the body. The individual pH in an organ is produced by its secretion of acids or alkalis which will establish the optimum level. The highly acid pH of the gastric juice is maintained by the secretion into the stomach of hydrochloric acid produced by chemical changes which take place in special cells in the walls of the stomach. The low pH value in the stomach provides the environment best suited to the functioning of the enzyme pepsin which is present in gastric juice. Saliva has a pH of between 6·4 and 6·8, which is the optimum value for the action of ptyalin.

Ptyalin action is inhibited when food containing it reaches the stomach and is mixed with highly acid gastric juice.

It will be noted that the blood has a pH value of between 7·4 and 7·5. This, therefore, is the general pH level in the body which may be altered in an individual organ such as the stomach, to meet the needs of the special functions of that organ. The range of pH of the blood which is compatible with life is 7·0 to 7·8. The metabolic activity of the body cells produces certain acids and alkalis which tend to alter the pH of the tissue fluid and the blood. To maintain the pH within the normal range there are substances present in the blood which act as *buffers*.

Buffers

Buffers are chemical substances such as phosphates, bicarbonates and some proteins, which are able to 'bind' free hydrogen ion or hydroxyl ion without change in pH. For example, if there is sodium hydroxide (NaOH) and carbonic acid (H_2CO_3) present, both will ionise to some extent but they will also react together to form sodium bicarbonate ($NaHCO_3$) and water (H.OH). One of the hydrogen ions from the acid has been 'bound' in the formation of the bicarbonate radical and the other by combining with the hydroxyl radical to form water.

The ability of the complex buffer systems of the blood to 'bind' hydrogen ions or neutralise acids is called the *alkali reserve* of the blood. If the pH of the blood falls below 7·4 the reserve of alkali has been reduced by an increase in production of hydrogen ions by the cells, and the condition of *acidosis* exists. When the reverse situation pertains and the pH is raised above 7·5 the increased alkali produced uses up the *acid reserve* and a state of *alkalosis* exists.

The buffer system serves to prevent dramatic changes in the pH values in the blood but it can only function effectively if there is some means by which excess acid or alkali can be excreted from the body. The two organs most active in this way are the lungs and the kidneys. When respiration is decreased there is an accumulation of carbon dioxide in the body which uses up the alkali reserve of the blood resulting in the development of acidosis. On the other hand if there is 'over-breathing' the condition of alkalosis may develop.

The kidneys have the ability to form ammonia which combines with the acid products of protein metabolism which are then excreted in the urine.

The buffer systems and the excretory systems of the body together maintain the *acid-base balance* of the body in health, so that the pH range of the blood remains within normal limits. If a pathological condition develops which causes a degree of alkalosis or acidosis a measure of the severity of the condition can be obtained by measuring the level of the acid or alkali reserve of the blood.

Body Fluids

The human body is made up of approximately 70 per cent of water, that is, a man weighing 70 kg consists of about 49 litres of water. Approximately 32 litres (65 per cent) of this water is intracellular, that is,

Figure 106

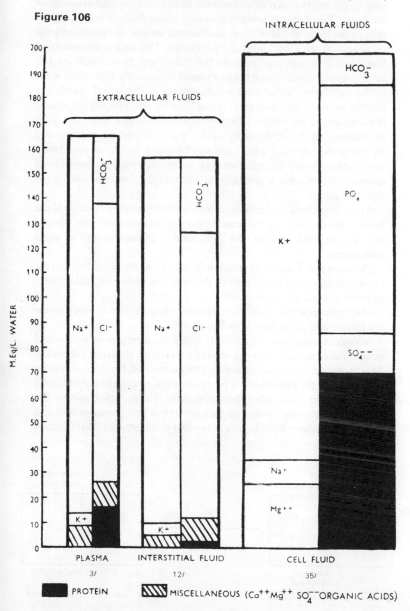

Fluid and electrolyte distribution in the body.

within the cells of the body and about 17 litres (35 per cent) is extra-cellular. The water in the body provides the medium for transport of many chemical substances dissolved in it.

The extracellular fluid consists of fluid in the blood and lymph vessels, the cerebro spinal fluid and the fluid in the interstitial spaces of the body. This latter is sometimes called tissue fluid. It bathes all the cells of the body and is in close association, on the one hand with the cells, and on the other, with the capillaries. The only constituents of blood which do not pass into the tissue fluid are those which are too large in size to pass through the tiny spaces between the cells of the walls of the capillaries. Substances which remain in the blood vessels are plasma proteins, erythrocytes, thrombocytes and leucocytes, except those which are amoeboid. Electrolytes and nutritional materials, such as glucose, amino acids, fatty acids, glycerol, carbon dioxide and oxygen are all in units which are small enough in size to pass freely across the walls of the capillaries. This means that the concentration of electrolytes in plasma is virtually the same as the concentration in the tissue fluid.

As the tissue fluid provides the external environment of the cells of the body, measurement of the electrolyte concentration in the plasma provides an indication of the intracellular concentrations of these substances.

The same easy transfer of small molecules from the tissue fluid across the cell membrane into the cells of the body does not exist. Why this is so is not completely understood.

The concentrations of different electrolytes in the two compartments are known and are indicated in Figure 106.

In each compartment of the body fluids the numbers of anions and cations balance each other maintaining a state of electrical neutrality. This, however, is a dynamic state. The number of ions is continually changing and the physiological processes work to maintain the balance between anions and cations. Any substantial change, which may be due to disease processes, upsets the balance in one compartment first and this is reflected later by changes in the other compartment. Cell function

Figure 107

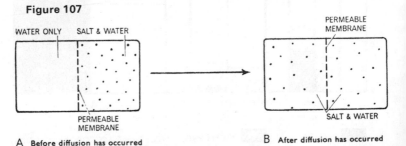

A   Before diffusion has occurred

B   After diffusion has occurred

Schematic diagram of the process of diffusion.

is dependent on electrolyte balance, not only inside the cell, but also in the environmental interstitial fluid.

Before going on to discuss the physiological factors which maintain fluid and electrolyte balance in the body, there are two physical processes which require explanation. These are *diffusion and osmosis*, both of which involve the movement of substances across semi-permeable membranes.

A semi-permeable membrane is a membrane which has 'pores' in it which allow for the passage of substances which are small enough in size. Although weight is not always an indication of the size and shape of a molecule, molecular weight can be used as a guide to the relative sizes of chemical substances.

Diffusion

Diffusion is the physical process of the transfer of dissolved substances across a membrane in order to establish equality of concentration on the two sides of the membrane. Diffusion is a fairly slow process and it takes time for this equality to occur. This type of movement of dissolved substances is not the deliberate movement of a specific number of particles (as shown in the illustration) but results from the constant movement which is characteristic of all substances in solution. The particles, in their constant movement, bombard the dividing membrane and some pass through. The net result of this bombardment is equality of concentration of substances on the two sides of the membrane.

Osmosis

Osmosis is the process of the transfer of the *water* of a solution across a semi-permeable membrane and the force with which this occurs is called the *osmotic pressure*.

Figure 108

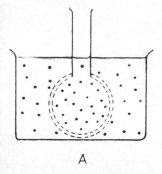

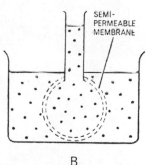

A   B

Schematic diagram of the process of osmosis.

It would be easier to understand this phenomenon if osmotic pressure was considered as *osmotic pull*, because water crosses the semi-permeable membrane from the low concentration side to the high concentration side. By taking water away from the side of lower concentration its concentration increases and by adding it to the high concentration solution its concentration is reduced. The process will continue until the concentrations on each side of the membrane are equal; when this happens they are then said to be *isotonic*. An example may help to clarify this. If a dilute solution of gelatine and water is prepared and placed in a beaker and a concentrated solution placed inside a semi-permeable membrane, such as a cellophane bag or sausage skin, as shown in Figure 108, water passes from the solution of lower concentration to the solution of higher concentration. The osmotic pull which is exerted is directly related to the difference in the concentration of the dissolved substances on the two sides of the membrane.

The processes of diffusion and osmosis have been described separately, but it must be understood that within the body the two processes occur concurrently.

Osmosis and diffusion cannot explain the transfer of all substances which are known to cross living cell membranes. Two further explanations offered for the transfer of such substances are:

1. That some substances can dissolve in the fat of the cell membrane and diffuse through it until they enter the cytoplasm of the cell.

2. That some substances can be *actively transferred* through the walls of the cell. For each substance transferred in this way there must be a specific carrier substance A in the cell membrane. The chemical B then combines with substance A to make a solution in the cell membrane. The solution then diffuses through the cell membrane into the cytoplasm. Substance B is then released and the chemical carrier A returns to the cell membrane in preparation for the transfer of another molecule of substance B.

7. The Circulatory System

The circulatory or vascular system is divided for descriptive purposes into two main parts.

1. *The blood circulatory system* which consists of:
   the heart, which acts as a pump
   the blood vessels through which the blood circulates.

The heart and the blood vessels form a completely closed and continuous system through which the blood flows.

2. *The lymphatic system* which consists of:
   lymphatic glands
   lymphatic vessels and capillaries through which a colourless fluid known as lymph circulates.

The two systems communicate with one another and are intimately associated.

THE STRUCTURE OF THE BLOOD VESSELS

There are several types of blood vessels:
   arteries
   arterioles
   veins
   venules
   capillaries.

The arteries

These are the blood vessels which transport blood away from the heart. They vary considerably in size, but have much the same structure.

They consist of three layers of tissue:
   the *tunica adventitia* or outer layer consists of fibrous tissue
   the *tunica media* or middle layer consists of elastic and smooth muscular tissue
   the *tunica intima* or inner lining consists of squamous epithelium called endothelium.

Figure 109

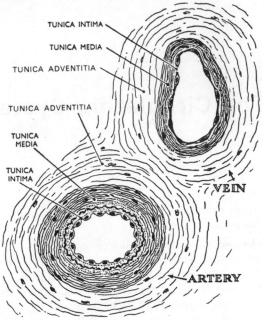

TUNICA INTIMA

TUNICA MEDIA

TUNICA ADVENTITIA

TUNICA ADVENTITIA

TUNICA MEDIA

TUNICA INTIMA

VEIN

ARTERY

Structure of an artery and a vein.

The amount of muscular tissue and elastic tissue varies in the arteries depending upon their size. In the large arteries the tunica media consists of more elastic tissue and less muscle. These proportions gradually change as the arteries become smaller until, in the arterioles, there is very little elastic tissue and a high proportion of smooth muscle.

The veins

The veins are the blood vessels which transport blood to the heart. The walls of the veins like those of the arteries have three layers of tissue:

tunica adventitia

tunica media

tunica intima.

The main difference between the walls of the veins and arteries is the comparative *thinness and weakness* of the tunica media of the veins. There is relatively less muscular and elastic tissue present. The walls of the veins are thin and collapse when cut while those of the arteries are thicker and retain their cylindrical shape when cut.

Some veins possess *valves* which prevent the back flow of blood. They are abundant in the veins of the limbs and are usually absent from the veins in the thorax and the abdomen. The valves are formed by a fold of tunica intima strengthened by connective tissue. They are *semi-lunar* in shape with the concavity towards the heart.

Figure 110

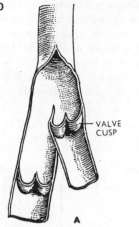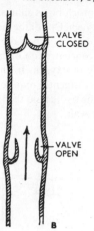

Interior of a vein.

(A)   Showing the valves.

(B)   Arrow showing direction of blood flow through a valve.

The capillaries

The small arteries known as the arterioles break up into a number of minute vessels known as *capillaries*. The wall of a capillary is composed of a single layer of endothelial cells which is very thin and permits the passage of water, substances of small molecular size and oxygen, but not the passage of red blood corpuscles or plasma proteins. Their diameter is approximately that of a red blood corpuscle.

THE NERVOUS CONTROL OF THE BLOOD VESSELS

Both the veins and arteries are supplied by nerves through the *autonomic nervous system*. These nerves arise from the *vaso-motor centre* in the *medulla oblongata*. The nerves from this centre are responsible for the changes in calibre of the vessels, thus controlling the amount of blood circulating to any part of the body at any point in time.

The nerves which reduce the lumen of the blood vessels are known as *vaso-constrictors* and those which increase the lumen, *vaso-dilators*.

CELL RESPIRATION

Internal or cell respiration is the name given to the interchange of gases between the blood and the cells of the body.

Oxygen is carried from the lungs to the tissues in chemical combination with haemoglobin, *as oxyhaemoglobin*. The exchange in the tissues takes place between the arterial end of the capillaries and the tissue fluid. The process involved is that of *diffusion from a higher concentration of oxygen in the blood to a lower concentration in the tissues*.

Oxyhaemoglobin is an unstable compound which breaks up easily to liberate oxygen. One of the factors which assists the liberation of oxygen is the amount of carbon dioxide present. In active tissues there is an increased concentration of carbon dioxide which leads to an increased availability of oxygen. In this way oxygen is available to the tissues which most urgently require it. Oxygen *diffuses* through the capillary wall into the tissue fluid then into the cell protoplasm through its semi-permeable wall.

Figure 111

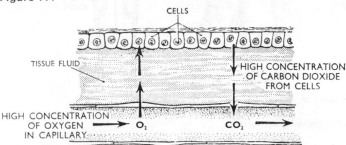

Scheme of internal or cell respiration.

Carbon dioxide is one of the waste products of carbohydrate and fat metabolism in the cells of the body. The mechanism of transfer of carbon dioxide from the cell into the blood, now at the *venous end of the capillary* is also by *diffusion*. Blood transports carbon dioxide to the lungs for excretion, by three different mechanisms:

some of the carbon dioxide is dissolved in the water of the blood plasma

some is transported in chemical combination with sodium in the form of sodium bicarbonate

the remainder is transported in combination with haemoglobin.

Cell Nutrition

The nutritive materials which the cells of the body require are transported round the body in the plasma of the circulating blood. In the process of passing from the blood to the cells the nutritive materials pass through the semi-permeable capillary walls into the tissue fluid which bathes the cells, then through the cell wall into the cell. The mechanism of the transfer of water and other substances from the blood capillaries depends mainly upon two physical principals:

diffusion

osmosis.

Diffusion

The walls of the blood capillaries consist of a single layer of endothelial cells which constitute a *semi-permeable membrane*. This membrane allows for the passage of *low molecular weight substances* through the capillary wall and the retention within the capillary of *high molecular weight substances*.

Nutrient materials which are in solution and are of low molecular weight pass through the semi-permeable membrane by *diffusion*, that is, they pass from a *high concentration in the blood* to a *lower concentration in the tissue fluid* and thus to the cells. Glucose, amino acids, fatty acids and glycerol, mineral salts, vitamins and water which are necessary for the formation and functioning of cells are all of relatively low molecular weight and pass out into the tissue fluid *by diffusion*.

Osmotic pressure

This is the pressure developed across a semi-permeable membrane which forces water to pass from a more dilute solution to a more concentrated solution, in an attempt to establish a state of equilibrium. The extent of the osmotic pressure depends upon the *number of non-diffusable particles* in solution on each side of the membrane.

Substances of low molecular weight can diffuse freely across a semi-permeable membrane and achieve the same concentration on each side of the membrane, therefore they are not involved in osmosis. The substances which influence osmotic pressure are those of molecular size too great to pass through the membrane. The *plasma proteins* in the capillaries are the substances mainly responsible for the osmotic pressure between the blood and tissue fluids.

At the arterial end of the capillaries the blood pressure is about *40 mm of mercury*. This is the pressure which tends to force substances into the tissue spaces. The osmotic pressure in the capillaries, exerted mainly by the plasma proteins is about *25 mm of mercury*, this tends to retain water within the blood vessels. The net outward pressure is the difference between these two, that is, *15 mm of mercury*.

Figure 112

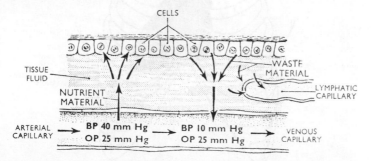

Scheme to illustrate interchange of nutrient material and waste material between capillaries and cells.

At the venous end of the capillaries the blood pressure drops to about *10 mm of mercury* while the osmotic pressure remains the same at *25 mm of mercury ;* thus there is a net pressure drawing fluid into the blood vessel of about *15 mm of mercury.*

This transfer of substances, including water, to the tissue spaces is a dynamic process. Blood flows through the capillaries from the arterial to the venous end and there is a constant change. All the water and cell waste products do not return to the venous capillaries. The excess is drained away from the tissue spaces in the minute *lymphatic capillaries.*

The tiny lymphatic capillaries originate as blind end tubes with walls similar to, but more permeable than, those of the blood capillaries. Extra tissue fluid and some of the cell waste enter the lymphatic capillaries and are eventually returned to the blood stream (see Chapter 8 on The Lymphatic System).

The Heart

The heart is a hollow muscular organ which is roughly cone shaped. It lies in the middle mediastinum between the lungs in the thoracic cavity. It lies obliquely, a little more to the left than the right and it presents a *base* above and an *apex* below. The heart is approximately 10 cm (4 inches) in length, weighs approximately 270 g (9 oz.), and is about the size of the owner's fist.

Figure 113

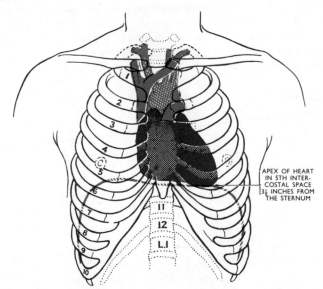

APEX OF HEART
IN 5TH INTER-
COSTAL SPACE
3½ INCHES FROM
THE STERNUM

The position of the heart.

Figure 114

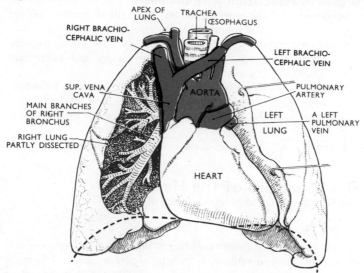

Diagram showing the organs in association with the heart.

Position of the heart

The heart lies behind the sternum and the *apex* lies about 9 cm (3½ inches) to the left of the sternum in the *fifth intercostal space* on the *midclavicular line*. The *base* extends to the *second costal cartilage* about 1·5 cm (½ inch) to the left of the sternum.

Figure 115

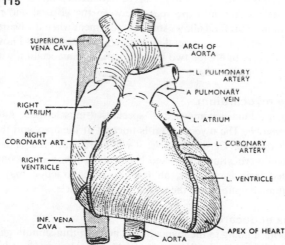

The heart, showing the position of the atria and ventricles.

Organs in association with the heart

Inferiorly: the apex rests on the central tendon of the diaphragm.

Superiorly: the great blood vessels, i.e., the aorta, superior vena cava, pulmonary artery and pulmonary veins.

Posteriorly: the oesophagus, trachea, left and right bronchus, descending aorta and thoracic vertebrae.

Laterally: the lungs. The left lung overlaps the left side of the heart.

Anteriorly: the sternum, the ribs and the intercostal muscles.

The Structure of the Heart

The heart is composed of several layers of tissue.

1. The pericardium

The pericardium is the outer covering of the heart and consists of two layers of tissue, the outer layer is made up of *fibrous tissue* and the inner of *serous membrane*. The serous membrane is reflected on to the surface of the heart muscle.

The *outer fibrous tissue* is continuous with the tunica adventitia of the great blood vessels above, and below it is attached to the central tendon of the diaphragm. Because of its fibrous nature it prevents over-distension of the heart.

The *serous membrane* lining the fibrous sac is known as the *parietal pericardium* and the *visceral pericardium* is that part directly attached to the heart muscle.

The serous membrane is made up of flattened epithelial cells which secrete serous fluid into the space between the visceral and parietal pericardium. This fluid allows smooth movement when the heart beats. The space between the parietal and visceral pericardium is known as a *potential space*. In life the two layers are in close association with only serous fluid between them.

2. The myocardium

The myocardium is composed of specialised muscle tissue known as *cardiac muscle*. The myocardium is thickest at the apex and thins out towards the base. Cardiac muscle, it will be remembered, is described as a pseudo-syncytium mass, and because the fibres are incompletely separated from each other, an impulse of contraction may spread without interruption over the whole sheet of muscle.

3. The endocardium

This forms a lining to the myocardium and is a thin smooth glistening membrane consisting of flattened epithelial cells known as *endothelium* which is continuous with the lining of the blood vessels.

Figure 116

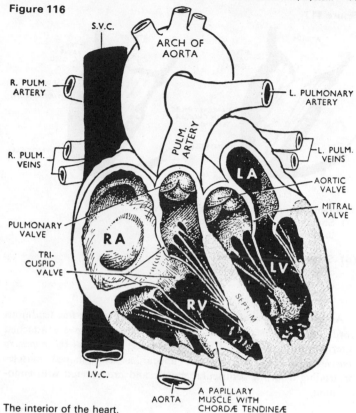

S.V.C.

ARCH OF
AORTA

R. PULM.
ARTERY

L. PULMONARY
ARTERY

PULM.
ARTERY

R. PULM.
VEINS

L. PULM.
VEINS

LA

AORTIC
VALVE

MITRAL
VALVE

PULMONARY
VALVE

RA

LV

TRI-
CUSPID
VALVE

RV

SEPTUM

I.V.C.

AORTA

A PAPILLARY
MUSCLE WITH
CHORDÆ TENDINEÆ

The interior of the heart.

INTERIOR OF THE HEART

The heart is divided into a right and left side by a partition of muscular
tissue known as the *septum*. The right and left sides are divided in turn
by partitions known as *valves* into upper and lower chambers. The upper
chambers are known as *atria*. The lower chambers are known as
ventricles. Therefore the heart has four chambers:

    right and left atrium

    right and left ventricle.

There is no communication between the right and left side of the heart
after birth.

The walls of the atria are thinner than those of the ventricles.

The valves dividing the atria from the ventricles are formed by
double folds of endocardium strengthened with fibrous tissue. The
valve separating the *right atrium* from the *right ventricle* is known as the
right atrioventricular valve, or *tricuspid valve*, and is made up of three
flaps or cusps. The valve separating the *left atrium* from the *left ventricle*
is termed the *left atrioventricular valve* or *mitral valve*, and is composed
of two flaps or cusps.

Figure 117

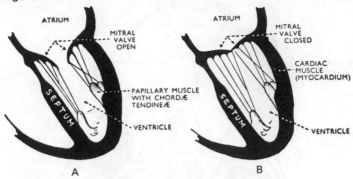

Diagram showing positions of the mitral valve.
(A) Showing the mitral valve open
to allow the left ventricle to fill.

(B) Showing the mitral valve
closed to prevent backflow
into the left atrium.

Attached to the ventricular surface of the valves are fine tendinous cords—the *chordae tendineae*. The chordae tendineae are attached inferiorly to projections of the heart muscle known as the *papillary muscles*. The papillary muscles appear as cone-shaped muscles protruding from the ventricular muscle and are covered with endocardium.

Figure 118

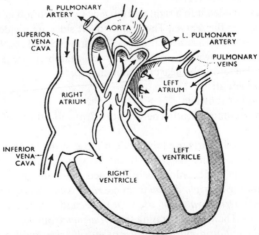

Diagram illustrating the flow of blood through the heart.

The Function of the Valves. Blood flows from the atria to the ventricles. When the ventricles contract the valves close and the chordae tendineae support the valves preventing regurgitation of blood into the atria.

FLOW OF BLOOD THROUGH THE HEART

The two largest veins of the body, the superior and inferior venae cavae, empty their contents into the right atrium. This blood passes via the right atrioventricular valve into the right ventricle, and from the right ventricle it is pumped into the *pulmonary artery or trunk* (the only artery in the body to carry venous blood). The opening of the pulmonary artery is guarded by a valve known as the *pulmonary valve* and is formed by three *semi-lunar cusps*. This valve prevents the back flow of blood into the right ventricle when the ventricular muscle relaxes. The pulmonary artery passes through the wall of the heart and divides into a *left and right pulmonary artery*. These arteries carry the venous blood to the lungs where an interchange of gases occurs; carbon dioxide being given off and oxygen taken up.

Figure 119

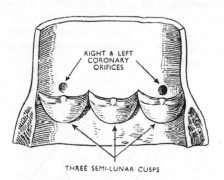

Diagram showing the cusps of a semi-lunar valve.

The arterial blood is carried from each lung by *two* pulmonary veins and the *four pulmonary veins* empty their contents into the left atrium of the heart. This blood now passes via the left atrioventricular valve into the left ventricle. From the left ventricle the arterial blood is now pumped into the aorta, the largest artery in the body, and subsequently into the general circulation. The opening of the aorta is guarded by a valve termed the *aortic valve* which is formed by three *semi-lunar cusps*.

The walls of the atria are thinner than the walls of the ventricles. The atria merely pump the blood through the atrioventricular valves into the ventricles. It is to be noted that both atria *contract simultaneously*. The ventricles have to pump the blood to the lungs via the right ventricle and into the general circulation via the left ventricle. The myocardium of the left ventricle is therefore much thicker and stronger than that of the right ventricle. The right and left ventricle contract *at the same time*.

The pulmonary trunk leaves the heart from the upper part of the right ventricle, and the aorta leaves from the upper part of the left ventricle.

Note

The *right side* of the heart deals with the flow of *venous blood*.
The *left side* of the heart deals with the flow of *arterial blood*.
The vessels *carrying blood to* the heart are *veins*.
The vessels *carrying blood away from* the heart are *arteries*.

THE CONDUCTING SYSTEM OF THE HEART

Situated within the heart muscle are small groups of specialised neuro-muscular cells which provide the system for conducting the impulses of contraction of the cardiac muscle.

The sinuatrial node

This small mass of specialised cells is situated in the wall of the right atrium near the opening of the superior vena cava. The sinuatrial node is often described as the '*pace-maker*' of the heart because it is capable of initiating impulses which stimulate the myocardium to contract. Without any outside influence from the nervous system it can stimulate the heart to beat about 60 times per minute.

Figure 120

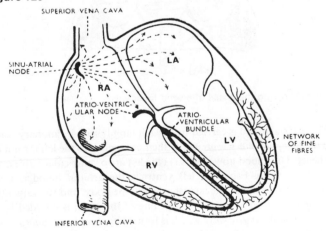

Diagram illustrating the conducting system of the heart.

The atrioventricular node

This mass of neuro-muscular tissue is situated near to the septum of the heart close to the atrioventricular valves. Normally the atrioventricular node is stimulated by the contraction of atrial myocardium, however, it is capable of initiating impulses of contraction at the rate of about 45 per minute.

The atrioventricular bundle

This consists of a mass of specialised fibres which originate from the atrioventricular node and passes downwards in the septum which separates the right and left ventricles. This bundle of fibres then divides into two branches, one going to each ventricle. Within the myocardium of the ventricles the branches break up into a network of fine filaments or fibres known as the fibres of Purkinje. The atrioventricular bundle and the Purkinje fibres convey the impulse of contraction from the atrioventricular node to the myocardium of the ventricles.

BLOOD SUPPLY TO THE HEART

The heart is supplied with arterial blood by the *coronary arteries*. There is a right and left coronary artery and they branch off the aorta immediately distal to the aortic valve.

The venous blood passes straight into the right atrium by the *coronary sinus*.

THE FUNCTION OF THE HEART

The function of the heart is to maintain a constant circulation of blood throughout the body. The heart acts as a pump and its action consists of a series of events known as the *cardiac cycle*.

The cardiac cycle

In a human being, when the heart is beating normally the cardiac cycle occurs about 74 times per minute. Thus each cycle takes about *0·8 of a second* to occur.

Figure 121

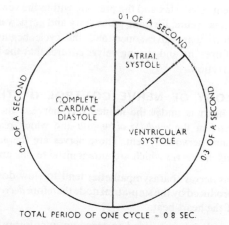

TOTAL PERIOD OF ONE CYCLE = 0 8 SEC.

Diagram illustrating the period of one cardiac cycle.

The *cardiac cycle* consists of:

atrial systole or contraction of the atria

ventricular systole or contraction of the ventricles

complete cardiac diastole, or relaxation of the atria and ventricles.

It does not matter at which stage of the cardiac cycle a description starts. For convenience the period when the atria are filling has been chosen.

The large veins, the superior vena cava and the inferior vena cava, pour venous blood into the right atrium, *at the same time* as the four pulmonary veins pour arterial blood into the left atrium. The sinu-atrial node emits an impulse of contraction which spreads as a wave of contraction over the atrial myocardium pushing the blood through the atrioventricular valves into the ventricles (atrial systole 0·1 sec). When this wave of contraction reaches the atrioventricular node it is stimulated to emit an impulse of contraction which spreads to the ventricular muscle via the atrioventricular bundle and the Purkinje fibres. This results in a wave of contraction which sweeps upwards from the apex of the heart and pushes the blood into the pulmonary artery and the aorta (ventricular systole 0·3 sec).

After contraction of the ventricles the heart rests for *0·4 of a second,* and this period is known as *complete cardiac diastole.* After complete cardiac diastole the cycle begins again with atrial systole.

The valves of the heart and of the great vessels open and close according to the pressure within the chambers of the heart. The atrio-ventricular valves are open while the ventricular muscle is relaxed during atrial systole. When the ventricles contract there is a gradual increase in the pressure in these chambers and when it rises above atrial pressure the atrioventricular valves close. When the ventricular pressure rises above that in the pulmonary artery and in the aorta, the pulmonary and aortic valves open and blood flows into these vessels. When the ventricles relax and the pressure within the ventricles falls the reverse process occurs. First the pulmonary and aortic valves close then the atrioventricular valves open and the cycle begins again. This sequence of opening and closing valves ensures that the blood flows in only one direction.

PHYSIOLOGY OF NERVE CONTROL OF THE HEART

In life the heart is under the influence of nerves which arise in the *cardiac centre* in the *medulla oblongata* and which reach it through the autonomic nervous system. These nerves are the *parasympathetic* and *sympathetic nerves* which are antagonistic to one another.

The *vagus nerves* (parasympathetic) tend to slow down the rate of impulses produced by the sinu-atrial node therefore *decreasing* the force and rate of the heart beat.

The *sympathetic nerves* tend to speed up impulses produced by the sinu-atrial node thus *increasing* the rate and force of the heart beat.

If stimuli are being received equally at the sinu-atrial node the normal heart contraction or beat will be maintained.

The nerve control can be demonstrated by the fact that during exercise sympathetic stimulation predominates, thus the rate and force of the heart beat is increased.

HEART SOUNDS

The individual is not usually conscious of his or her heart beat, but if the ear or the receiver of a stethoscope is placed upon the chest at about the midclavicular line in the fifth intercostal space the beat of the heart can be heard.

Two sounds, which are separated by a short pause, can be clearly distinguished. They are described in words as *'lubb dup'*. The first sound *'lubb'* is fairly loud and is due to:

the contraction of the ventricular muscle

the closure of the atrio-ventricular valves.

The second sound *'dup'* is softer and is due to the closure of the aortic and pulmonary valves after the ejection of blood through them.

Figure 122

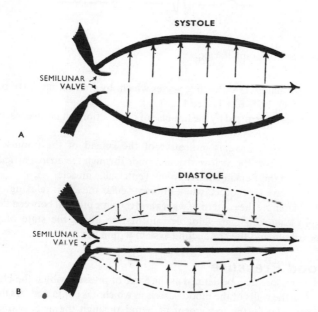

Diagram illustrating elasticity of walls of the aorta.

(A) Showing expansion during ventricular systole. Blood flow⟶

(B) Showing relaxation during cardiac diastole. Blood flow⟶

ELECTRICAL CHANGES IN THE HEART

When muscles contract there is a change in the electrical potential across the membrane of muscle fibres. As the body fluids and tissues are good conductors of electricity the electrical changes which occur in the contracting myocardium can be detected by attaching metal conductors to the surface of the body. The pattern of electrical change may be displayed on an oscilloscope screen or printed out on paper. This tracing is called an *electrocardiogram* (ECG).

The parts of the body which are usually 'tapped' are the arms, legs and chest. There are a number of standard combinations of attachments for tapping the electrical changes which occur which are called *leads*.

Lead I: right hand or arm and left hand or arm.
Lead II: right hand or arm and left foot or leg.
Lead III: left hand or arm and left foot or leg.
Lead IV: left foot or ankle and chest, over the heart.

Figure 123

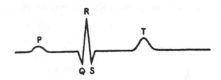

An electrocardiograph tracing.

The normal ECG shows five waves which, by convention, have been named P, Q, R, S and T, (see Fig. 123).

The P wave is caused by the impulse of contraction which sweeps over the atria.

The Q, R, S wave is indicative of the spread of the impulse of contraction from the atrioventricular node through the atrioventricular bundle and the Purkinje fibres to the ventricular muscle.

The T wave is recorded while the ventricular muscle is relaxing.

By examining the pattern of waves and the time interval between them the physician can gain valuable information about the state of the myocardium and the conducting system within the heart.

Blood Pressure

Blood pressure may be defined as the force or pressure which the blood exerts on the walls of the blood vessels in which it is contained. As there is some delay in the movement of blood through the arteriolar and capillary systems the blood pressure in the arteries is higher than that in the veins. This means that the arteries are always full and their walls are continuously subjected to stretch.

The arterial blood pressure is the result of the discharge of blood from the left ventricle into the *already full aorta*.

When the left ventricle contracts and pushes blood into the aorta the pressure produced is known as the *systolic blood pressure* which is found in an adult to be about *120 mmHg* (millimetres of mercury).

When complete *cardiac diastole* occurs and the heart is resting with no ejection of blood, the pressure within the blood vessels is termed the *diastolic blood pressure* which is found in an adult to be about *80 mmHg*.

The blood pressure is measured by the use of a sphygmomanometer and is usually expressed in the following manner:

$$BP = \frac{120}{80} \text{ mmHg}.$$

MAINTENANCE OF NORMAL BLOOD PRESSURE

A number of factors are involved in the maintenance of the blood pressure:

the cardiac output
the blood volume
the peripheral resistance
the elasticity of the arterial walls
the venous return.

Cardiac output

The cardiac output may be considered as the amount of blood ejected from the heart by each contraction of the ventricles (stroke volume), or the amount ejected each minute (minute volume). The minute volume takes into consideration the rate *and* force of cardiac contraction. An increase in minute volume raises both the systolic and diastolic pressure but an increase in the stroke volume increases the systolic pressure more than it does the diastolic pressure.

Blood volume

A sufficient amount of blood must be circulating in the vessels to maintain the normal blood pressure. If, for example, a haemorrhage has occurred and a large amount of blood is lost, the volume is reduced with subsequent lowering of the pressure.

The peripheral or arteriolar resistance

This refers to the small blood vessels; the arterioles and the capillaries but particularly the arterioles. The arterioles are extremely narrow tubes and subsequently offer a resistance to the flow of blood. This resistance prevents the too speedy flow of blood into the capillaries and thus assists in maintaining the blood pressure within the arterial system.

If the arterioles are stimulated by the vaso-constrictors their lumen will be reduced which will increase the resistance to the blood flow, thus raising the blood pressure. On the other hand if the arterioles are stimulated by the vaso-dilators then their lumen will be increased and more blood will flow through to the capillaries with a subsequent lowering of blood pressure.

The elasticity of the arterial walls

There is a considerable amount of elastic tissue in the arterial walls. Therefore, when the left ventricle ejects blood into the already full aorta it distends and then recoils pushing the blood onwards. This distension occurs all through the arterial system. During cardiac diastole the vessels recoil and press upon the blood pushing it onwards through the arterioles, thus maintaining the diastolic pressure.

Venous return

The amount of blood returned to the heart through the superior and inferior venae cavae plays an important part in cardiac output. The force of contraction of the left ventricle ejecting blood into the aorta and subsequently into the arteries, arterioles and capillaries is not sufficient to return the blood through the veins back to the heart, therefore, other factors are involved in assisting the venous return.

The Position of the Body. Gravity assists the venous return from the head and neck when the individual is standing or sitting.

Muscular Contraction. The contraction of muscles, particularly skeletal muscles, puts pressure on the veins. This squeezing or milking action has the effect of pushing the blood towards the heart. Backward flow of blood is prevented by the valves in the veins.

Effects of Respiratory Movements. During inspiration the expansion of the chest creates a negative pressure within the thorax. This has the effect of assisting the flow of blood towards the heart. In addition when

Figure 124

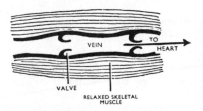

Muscle relaxed, valves open.

Muscle contracted squeezing vein, valve nearest heart open to allow flow of blood towards heart.

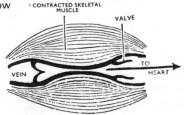

Diagram illustrating the flow of blood through a vein controlled by muscle contraction.

the diaphragm descends during inspiration the intra-abdominal pressure rises, and this squeezes blood towards the heart.

The Pulse

The pulse is described as a wave of distension and elongation felt in an artery wall due to the contraction of the left ventricle forcing about 70 millilitres of blood into the already full aorta. When the aorta is distended a wave passes along the walls of the arteries and can be felt at any point where an artery can be pressed gently against a bone. The waves occur approximately 74 times per minute in health and represents the number of heart beats.

It must be appreciated that this wave is felt in the artery wall before the blood ejected into the aorta could possibly reach the area. The wave travels about ten to fifteen times more rapidly than does the blood itself and is quite independent of it.

Certain points must be noted when recording the pulse of an individual.

The rate at which the heart is beating.

The rhythm, that is, the regularity with which the heart beats occur, the length of time between each beat should be the same.

The volume or strength of the beat. It should be possible to obliterate the artery with moderate pressure.

The tension. The artery should feel soft and pliant under the fingers, not hard and tortuous.

Factors affecting the pulse rate

Position. When the individual is standing up the pulse rate will be more rapid than when he is lying down.

Age. The pulse rate in children is more rapid than in adults.

Sex. The pulse rate tends to be more rapid in the female than in the male. The difference is usually about five beats per minute.

Exercise. Any exercise, walking, running or playing games will increase the rate of the pulse.

Emotion. When any strong emotion is experienced the pulse rate is increased, for example, excitement, fear, anger, grief.

The Circulation of the Blood

Three systems are discussed when considering the circulation of blood throughout the body:

  the pulmonary circulation
  the systemic or general circulation
  the portal circulation.

Figure 125

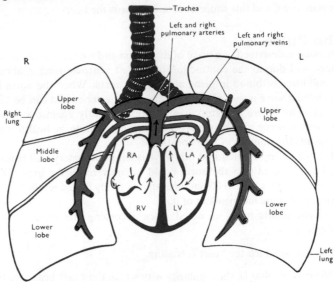

Schematic diagram of pulmonary circulation.

THE PULMONARY CIRCULATION

The pulmonary circulation consists of the circulation of blood from the right ventricle of the heart to the lungs and back to the left atrium, during which carbon dioxide is excreted and oxygen is absorbed.

The pulmonary artery or trunk, carrying deoxygenated blood, leaves the upper part of the right ventricle of the heart. It passes upwards through the walls of the heart and at the level of the fifth thoracic vertebra it divides into a left and right pulmonary artery.

The left pulmonary artery runs to the root of the left lung where it divides into two branches one passing into each lobe.

The right pulmonary artery passes to the root of the right lung and divides into two branches. The larger branch carrying blood to the middle and lower lobes, and the smaller branch to the upper lobe.

Within the lung these arteries divide and subdivide into smaller arteries, subsequently becoming arterioles and capillaries. It is between the capillaries and the lung tissue that the interchange of gases occurs. The capillaries now containing oxygenated blood join up with one another and eventually form two veins.

Two pulmonary veins leave each lung, therefore *four* pulmonary veins, carrying oxygenated blood, enter the left atrium of the heart. During atrial systole this blood passes into the left ventricle, and during ventricular systole it is forced into the aorta and the general circulation.

Figure 126

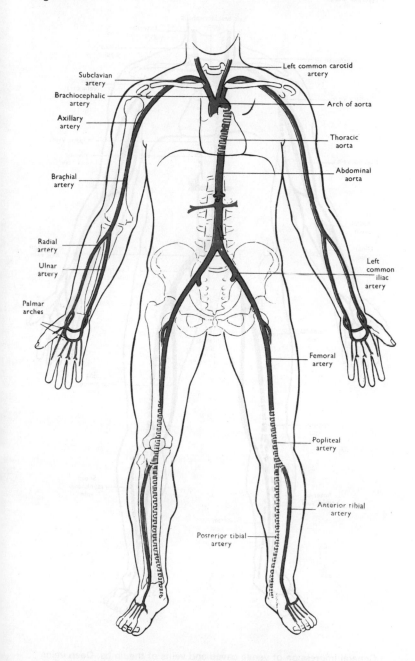

General impression of the aorta and main arteries of the limbs.

Figure 127

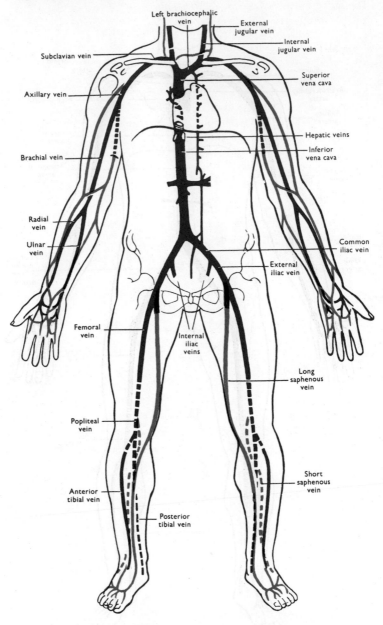

General impression of venae cavae and veins of the limbs. Deep veins in black and superficial veins in light blue.

THE GENERAL OR SYSTEMIC CIRCULATION

The blood pumped out from the left ventricle is carried by the aorta and its branches to all parts of the body.

THE AORTA

The aorta begins at the upper part of the left ventricle and after passing upwards for a short distance it arches backwards and to the left. It then descends behind the heart through the thoracic cavity a little to the left of the thoracic vertebrae. At the level of the 12th thoracic vertebra it passes through the aortic opening of the diaphragm then downwards to the level of the 4th lumbar vertebra where it divides into the two common iliac arteries.

Throughout its length the aorta gives off numerous branches. Some of the branches are *paired*, i.e., there is a right and left branch of the same name; others are single or *unpaired* branches.

Figure 128

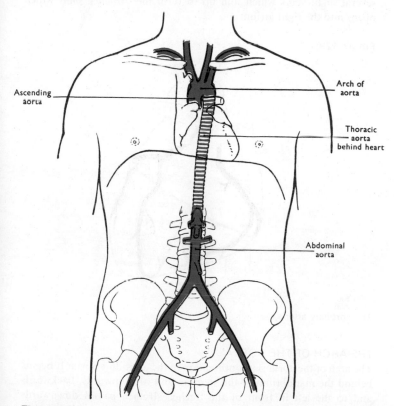

The aorta.

For descriptive purposes the aorta is divided into two parts:
the thoracic aorta
the abdominal aorta.

The thoracic aorta is described in three portions:
the ascending aorta
the arch of the aorta
the descending aorta.

THE ASCENDING AORTA

The ascending aorta is about 5 cm (2½ inches) in length and lies behind the sternum.

Branches of the ascending aorta

The Right and Left Coronary Arteries. These arteries supply the heart with oxygenated blood. As they traverse the heart they break up, eventually becoming capillaries. The venous blood is collected into several small veins which join up to form the *coronary sinus* which opens into the right atrium.

Figure 129

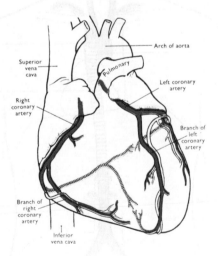

The coronary arteries.

THE ARCH OF THE AORTA

The arch of the aorta is a continuation of the ascending aorta. It begins behind the manubrium of the sternum and runs upwards, backwards and to the left in front of the trachea. It then passes downwards to the left of the trachea and is continuous with the descending aorta.

Figure 130

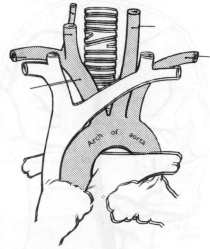

The arch of the aorta and its branches.

Branches of the arch of the aorta

Three important branches are given off from the upper aspect of the
arch of the aorta:

    the brachio-cephalic artery

    the left common carotid artery

    the left subclavian artery.

The Brachio-cephalic Artery. This artery is about 4 cm (2 inches) in
length and passes obliquely upwards, backwards and to the right. At
the level of the sterno-clavicular joint it divides into the *right common
carotid artery* and *right subclavian artery.*

CIRCULATION OF BLOOD TO THE HEAD AND NECK

The *right common carotid artery* is a branch of the brachio-cephalic
artery and the *left common carotid artery* arises directly from the arch of
the aorta. They pass upwards on either side of the neck and their
distribution on each side is identical. The common carotid arteries are
embedded in fascia, known as the *carotid sheath.* At the level of the
upper border of the thyroid cartilage they divide into:

    the external carotid artery

    the internal carotid artery.

Branches of the external carotid artery

This artery supplies the superficial tissues of the head and neck, giving
of a number of branches:

    The superior thyroid artery—supplying the thyroid gland.

Figure 131

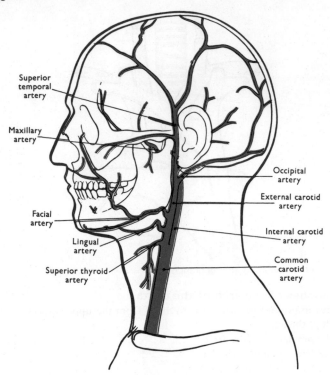

Superficial arteries supplying the face and head.

The lingual artery—supplying the tongue.

The facial artery—passes outwards over the mandible just in front of the angle of the jaw and supplies the muscles of facial expression.

The occipital artery—supplies the posterior part of the scalp.

The temporal artery—passes upwards over the zygomatic process in front of the ear and supplies the frontal, temporal and parietal parts of the scalp.

The maxillary artery—supplies the muscles of mastication, and a branch of this artery the *middle meningeal artery* runs deeply to supply structures in the interior of the skull.

Branches of the internal carotid artery

The internal carotid artery supplies the greater part of the brain, the eye, forehead and nose. It ascends to the base of the skull and passes through the carotid foramen in the temporal bone. Many branches arise from the internal carotid artery, *four* of these branches are:

the ophthalmic artery—supplying the eye with arterial blood
the anterior cerebral artery
the middle cerebral artery $\Big\}$ supplying the brain
the posterior communicating artery.

The greater part of the brain is supplied with arterial blood by a striking arrangement of arteries. This arrangement is known as the *circulus arteriosus* or *the circle of Willis*.

The circulus arteriosus or the circle of Willis

The *two anterior cerebral arteries* arise from the internal carotid arteries and are joined together by an artery known as the *anterior communicating artery*.

Figure 132

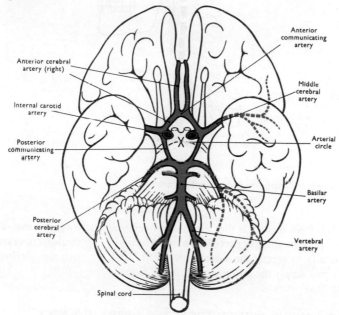

The arterial blood supply to the brain, showing the circulus arteriosus.

Posteriorly two *vertebral arteries* which rise from the subclavian arteries pass through the foramina in the transverse processes of the cervical vertebrae and enter the skull through the foramen magnum. Just inside the skull they join together to form the *basilar artery*. After travelling for a short distance the basilar artery divides to form the two *posterior cerebral arteries*. Each of these arteries is joined to the internal carotid arteries by the *posterior communicating arteries*. The *circulus arteriosus* is therefore formed by:

    2 anterior cerebral arteries
    1 anterior communicating artery
    2 posterior communicating arteries
    2 posterior cerebral arteries (see Fig. 132).

Figure 133

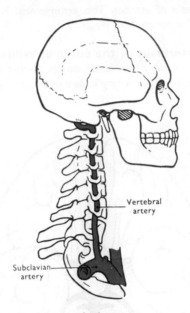

Diagram showing the right vertebral artery.

It is from this circle that the anterior cerebral arteries pass forwards to supply the anterior part of the brain. The middle cerebral arteries pass laterally to supply the sides of the brain, and the posterior cerebral arteries supply the posterior part of the brain.

THE VENOUS RETURN FROM THE HEAD AND NECK

The venous blood from the head and neck is returned mainly by *deep and superficial veins*.

It will be remembered that from the external carotid artery several branches supply arterial blood to the superficial parts of the face and scalp. *Superficial veins* with the same names return the venous blood from these areas and unite to form the external jugular vein.

The external jugular vein begins in the neck at the level of the angle of the jaw. It passes downwards in front of the sterno-mastoid muscle, then behind the clavicle to enter the *subclavian vein*.

The venous blood from the deep areas of the brain is collected into channels which are known as *sinuses*.

Figure 134

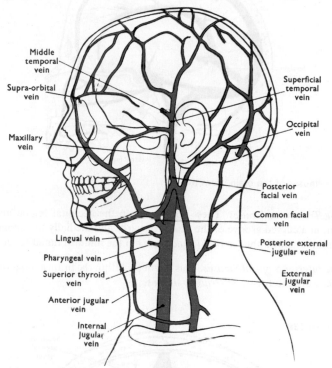

Middle temporal vein

Supra-orbital vein

Maxillary vein

Lingual vein

Pharyngeal vein

Superior thyroid vein

Anterior jugular vein

Internal jugular vein

Superficial temporal vein

Occipital vein

Posterior facial vein

Common facial vein

Posterior external jugular vein

External jugular vein

Venous return from the head and neck.

The venous sinuses of the brain

The walls of the venous sinuses are formed by layers of *dura mater* lined with endothelium. (The dura mater is the name given to the outer protective covering of the brain.) There are many venous sinuses, those most commonly described are:

1 superior sagittal sinus
1 inferior sagittal sinus
1 straight sinus
2 transverse or lateral sinuses (see Figs 135 and 136).

The Superior Sagittal Sinus. This sinus carries the venous blood from the superior part of the brain. It commences in the frontal region in folds of dura mater and passes directly backwards in the midline of the skull to the occipital region. The venous blood is then poured into one of the *transverse sinuses*.

Figure 135

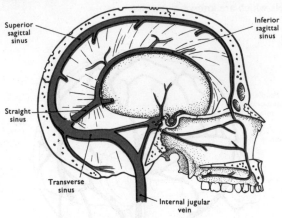

The sinuses of the brain, from the side.

The Transverse Sinuses. These commence in the occipital region and run, in a curved groove of the skull, forwards and medially to become continuous with the *internal jugular veins* in the middle cranial fossa.

The Inferior Sagittal Sinuses. This lies deep within the brain and passes backwards to form the *straight sinuses.*

Figure 136

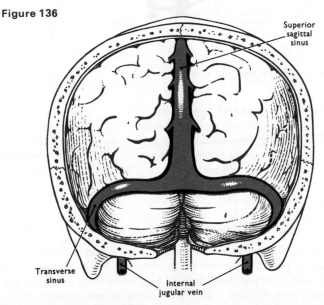

The sinuses of the brain, from above.

The Straight Sinuses. This runs backwards and downwards to become continuous with one of the transverse sinuses.

The Internal Jugular Veins. Each vein begins at the jugular foramen in the middle cranial fossa. It runs downwards in the neck behind the sterno-mastoid muscle. Behind the clavicle it units with the *subclavian vein* to form the *brachio-cephalic vein.*

The Brachio-cephalic Veins. These are two in number and are situated in the root of the neck. Each is formed by the internal jugular and the subclavian vein. The left brachio-cephalic vein is longer than the right and passes obliquely behind the manubrium of the sternum where it joins the right brachio-cephalic vein to form the *superior vena cava.*

The superior vena cava which drains all the venous blood from the head, neck and upper limbs, is about 7 cm (3 inches) in length. It passes downwards to the right of the sternum and ends in the right atrium of the heart.

CIRCULATION OF BLOOD TO THE UPPER LIMB

The subclavian arteries
The right subclavian artery arises from the brachio-cephalic artery; the left directly from the arch of the aorta. They are slightly arched and pass behind the clavicle and over the first rib before entering the axillae where they continue as the *axillary arteries.*

Before entering the axilla each subclavian artery gives off two branches.

Figure 137

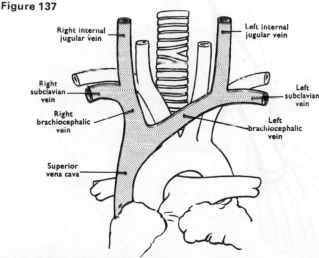

The formation of the superior vena cava.

The Vertebral Artery. This has already been described.

The Internal Mammary Artery. This supplies a number of structures in the thoracic cavity.

The Axillary Artery. This artery is a continuation of the subclavian artery and lies in the axilla. The first part lies deeply, then it runs more superficially to become the *brachial artery*.

Figure 138

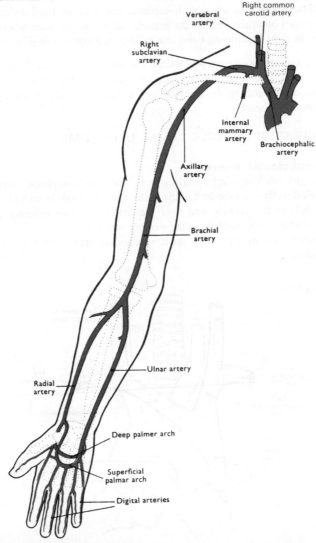

Arterial blood supply to the right arm.

The Brachial Artery. On leaving the axilla, the axillary artery continues as the brachial artery which runs down the medial aspect of the upper arm. It then gradually passes to the front and extends to about 1 cm below the elbow joint where it divides into the *radial and ulnar arteries.*

The Radial Artery. This passes down the radial side of the forearm to the wrist. In its lower part, just above the wrist it lies superficially and can be felt in front of the radius. It is here that the radial pulse is counted

Figure 139

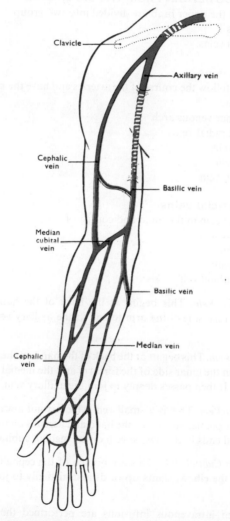

Clavicle

Axillary vein

Cephalic vein

Basilic vein

Median cubital vein

Basilic vein

Median vein

Cephalic vein

Venous return from the arm.

The Digital Arteries. These are arteries which arise from the palmar arches and supply the fingers.

The Ulnar Artery. This runs downwards on the ulnar or medial aspect of the forearm to the wrist.

Together the radial and ulnar arteries form two arterial arches in the hand; *the deep and superficial palmar arches.*

Branches from the axillary, brachial, radial and ulnar arteries supply all the structures in the upper limb.

THE VENOUS RETURN FROM THE UPPER LIMB

The veins of the upper limb are divided into two groups:

deep veins
superficial veins.

The deep veins

These veins follow the course of the arteries and have the same names:

digital veins
deep palmar venous arch
ulnar and radial veins
brachial vein
axillary vein
subclavian vein.

The superficial veins

These veins begin in the hand and consist of:

cephalic vein
basilic vein
median vein
median cubital vein.

The Cephalic Vein. This begins at the back of the hand and winds round the radial side of the arm to end in the axillary vein just below the clavicle.

The Basilic Vein. This begins at the back of the hand on the ulnar aspect. It ascends on the ulnar side of the forearm and the medial aspect of the upper arm. It then passes deeply to join the axillary vein.

The Median Vein. This is a small vein which is not always present. It begins at the palmar surface of the hand and ascends on the front of the forearm and ends in the basilic vein or the median cubital vein.

The Median Cubital Vein. This is a branch of the cephalic vein which, in front of the elbow, slants upwards and medially to join the basilic vein.

Note. When intravenous infusions are performed the vein chosen is commonly one of the superficial veins of the forearm or hand.

Brachio-cephalic Vein. This vein is formed when the subclavian and internal jugular veins unite.

Superior Vena Cava. This vein is formed when the two brachio-cephalic veins unite. It drains all the venous blood from the head, neck and upper limbs and terminates in the right atrium.

Figure 140

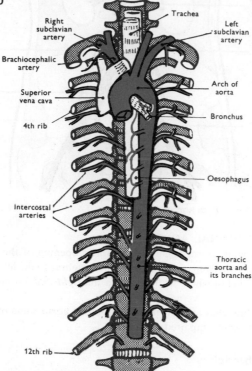

The arch of the aorta, thoracic aorta and their branches.

THE DESCENDING THORACIC AORTA

The descending thoracic aorta begins at the level of the 4th thoracic vertebra. It extends downwards in front of the anterior surface of the bodies of the thoracic vertebrae, to the level of the 12th thoracic vertebra where it passes through the aortic opening of the diaphragm to become continuous with the abdominal aorta.

The descending thoracic aorta gives off many branches which supply the walls of the thoracic cavity and the organs within the cavity. Some of these branches are:

bronchial arteries—supplying the bronchi
oesophageal arteries—supplying the oesophagus
intercostal arteries—supplying the intercostal muscles.

Figure 141

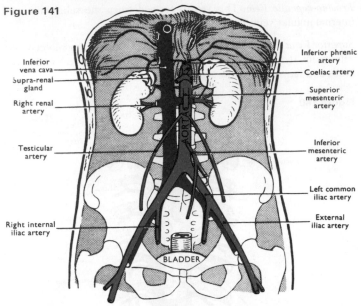

The abdominal aorta and its branches.

THE ABDOMINAL AORTA

The abdominal aorta begins at the aortic opening of the diaphragm and descends in front of the bodies of the vertebrae to the level of the 4th lumbar vertebra where it divides into the *right and left common iliac arteries.*

Many branches arise from the abdominal aorta some of which are paired and some unpaired.

Paired branches

Inferior Phrenic Arteries. These supply the diaphragm.

Renal Arteries. These supply the kidneys and give off branches to the supra-renal glands.

Testicular Arteries. These supply the testes in the male.

Ovarian Arteries. These supply the ovaries in the female.

Unpaired branches

Coeliac Artery. This is a short thick artery about 1·25 cm ($\frac{1}{2}$ inch) in length. It arises immediately below the aortic opening of the diaphragm. It divides into three branches:

left gastric artery—supplies the stomach
splenic artery—supplies the pancreas and the spleen
hepatic artery—supplies the liver and part of the stomach.

Superior Mesenteric Artery. This artery supplies the whole of the small intestine and the proximal half of the large intestine.

Inferior Mesenteric Artery. This artery arises from the aorta about 4 cm (1½ inches) above its division into the common iliac arteries. It supplies the distal half of the large intestine and part of the rectum.

Figure 142

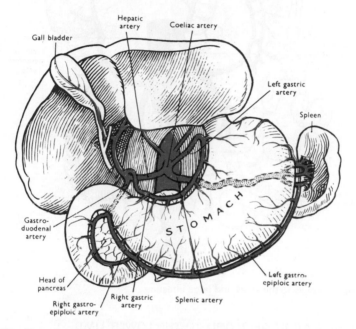

Blood supply to the stomach and spleen.

The Common Iliac Arteries. The right and left common iliac arteries are formed when the abdominal aorta divides at the level of the 4th lumbar vertebra. Each one divides in front of the sacro-iliac joints to form:

internal iliac artery
external iliac artery.

The *internal iliac artery* runs medially to supply the organs within the pelvic cavity. In the female one of the main branches is the *uterine artery* which supplies the uterus.

The *external iliac artery* is larger than the internal iliac artery. It runs obliquely downwards to the level of the inguinal ligament where it becomes the femoral artery.

Figure 143

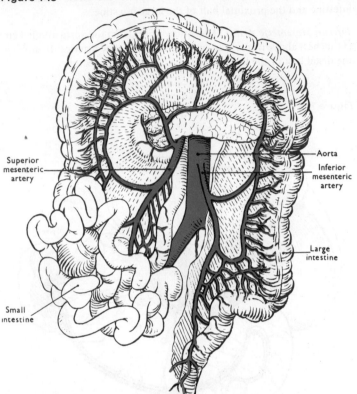

Superior mesenteric artery

Small intestine

Aorta

Inferior mesenteric artery

Large intestine

Blood supply to the small and large intestine.

CIRCULATION OF BLOOD TO THE LOWER LIMB

The Femoral Artery. This is a continuation of the external iliac artery. It begins behind the mid-point of the inguinal ligament and extends downwards, at first in front of the thigh, then medially, and eventually passes backwards behind the femur to enter the popliteal space. In the popliteal space it becomes the popliteal artery. It supplies blood to the structures of the thigh.

The Popliteal Artery. This passes through the popliteal space behind the knee. It supplies the structures in this space and the knee joint. At the lower border of the popliteal space it divides into the anterior and posterior tibial arteries.

The Anterior Tibial Artery. This passes forwards between the tibia and fibula, and supplies the structures in the front of the leg. It lies on the tibia and runs in front of the ankle joint and continues over the dorsum (front) of the foot as the dorsalis pedis artery.

The Posterior Tibial Artery. This runs downwards and medially on the back of the leg. In the lower part it becomes superficial and passes behind the medial malleolus to reach the sole of the foot where it continues as the plantar artery.

The Dorsalis Pedis Artery. This is a continuation of the anterior tibial artery and passes over the dorsum of the foot supplying arterial blood to the structures in this area.

The Plantar Artery. This supplies the structures in the sole of the foot. This artery and its branches form an arch similar to that of the hand, from which the digital branches arise to supply the toes.

Figure 144

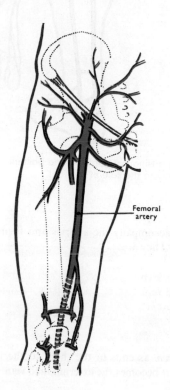

Femoral
artery

Diagram showing position of femoral artery.

THE VENOUS RETURN FROM THE LOWER LIMB

The veins returning venous blood from the lower limb are, like those of the upper limb, divided into deep and superficial veins.

Figure 145

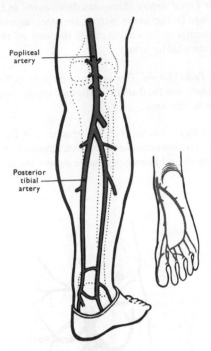

Popliteal
artery

Posterior
tibial
artery

Diagram showing position of popliteal and posterior tibial arteries. *Inset* shows blood supply to the sole of the foot.

Deep veins

The deep veins accompany the arteries and their branches and have the same names. They are:

> digital veins
> plantar venous arch
> posterior tibial vein
> anterior tibial vein
> popliteal vein
> femoral vein.

The femoral vein ascends in the thigh to the level of the inguinal ligament where it becomes the external iliac vein.

The superficial veins

The two main superficial veins draining the lower limbs are:

> short saphenous vein
> long saphenous vein.

The Short Saphenous Vein. This vein begins behind the lateral malleolus, being formed by the union of small veins which drain the dorsum of the foot. It ascends superficially along the back of the leg and in the popliteal space joins the *popliteal vein*.

The Long Saphenous Vein. This is the longest vein in the body. It begins at the medial aspect of the dorsum of the foot and runs upwards crossing the medial aspect of the tibia and up the inner side of the thigh. Just below the inguinal ligament it passes deeply to end in the *femoral vein*.

Figure 146

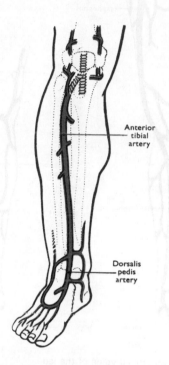

Anterior
tibial
artery

Dorsalis
pedis
artery

Diagram showing position of anterior tibial artery and blood supply to dorsum of foot.

THE VENOUS RETURN FROM THE PELVIC AND ABDOMINAL CAVITIES

The Internal Iliac Vein. This receives tributaries from several veins which drain the organs of the pelvic cavity. At the level of the sacro-iliac joint it units with the external iliac vein to form the common iliac vein.

The Two Common Iliac Veins. These begin at the level of the sacro-iliac joint. They ascend obliquely and end a little to the right of the body of the fifth lumbar vertebra by uniting to form the inferior vena cava.

Figure 147

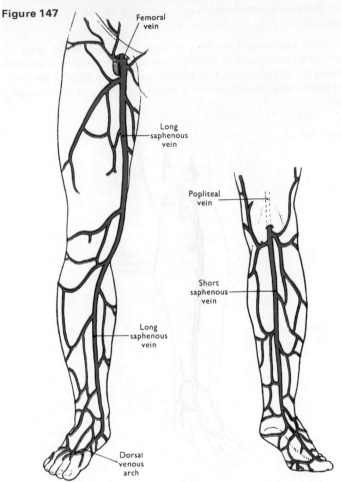

Diagram showing superficial veins of the leg.

The Inferior Vena Cava. This ascends on the right of the bodies of the vertebrae and conveys venous blood to the right atrium of the heart.

The veins which join the inferior vena cava in the abdominal cavity are:

testicular veins—in the male

ovarian veins—in the female

renal veins

right supra-renal vein (the left supra-renal vein drains into the left renal vein)

hepatic vein.

The inferior vena cava passes through the central tendon of the diaphragm at the level of the 8th thoracic vertebra and shortly after enters the right atrium of the heart.

VENOUS RETURN FROM THE THORACIC CAVITY

Most of the venous blood from the organs in the thoracic cavity is drained into the *azygos vein* and the *hemiazygos vein*. Some of the main veins which join these veins are the *bronchial, oesophageal* and *intercostal veins*. The azygos vein joins the superior vena cava and the hemiazygos vein joins the left brachio-cephalic vein.

Figure 148

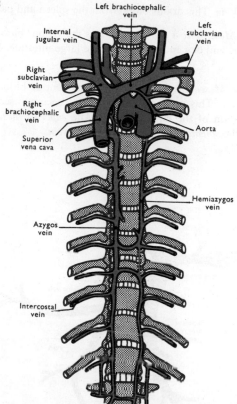

Diagram showing venous drainage of thoracic cavity into superior vena cava.

THE PORTAL CIRCULATION

In all the parts of the circulation which have been described previously venous blood has passed from the tissues to the heart by the most direct route. In the portal circulation blood passes from the abdominal part of the digestive system and the spleen via the liver and the inferior vena cava to the heart. In this way blood with a high concentration of nutrient materials goes to the liver first, where certain modifications take place including the regulation of their supply to other parts of the body.

The portal vein

This is formed by the following veins joining together:
  splenic vein
  inferior mesenteric vein
  superior mesenteric vein
  gastric veins
  cystic vein.

The Splenic Vein. This drains blood from the spleen and pancreas.

The Inferior Mesenteric Vein. This returns the venous blood from the rectum, pelvic and descending colon of the large intestine. It joins the splenic vein.

The Superior Mesenteric Vein. This returns venous blood from the small intestine. This vein also drains the caecum, ascending and transverse colon of the large intestine. The *superior mesenteric vein* unites with the *splenic vein* to form the *portal vein*.

Figure 149

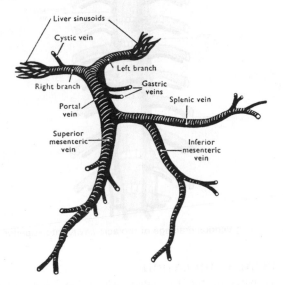

Schematic diagram of the portal vein.

The Gastric Veins. These veins from the stomach join the portal vein.

The Cystic Vein. This drains venous blood from the gall bladder and ends in the portal vein.

Figure 150

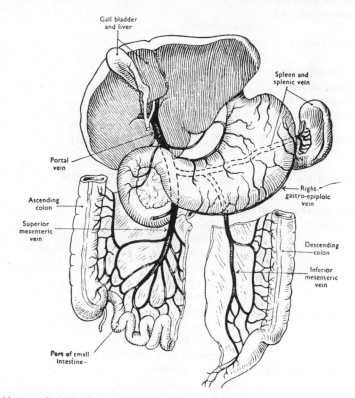

Venous drainage from some of the abdominal organs, veins joining to form the portal vein.

Summary of the Systemic Circulation

THE THORACIC AORTA

The ascending aorta

right and left coronary arteries.

The arch of the aorta

brachio-cephalic $\left\{\begin{array}{l}\text{right common carotid artery} \\ \text{right subclavian artery}\end{array}\right.$

left common carotid artery

left subclavian artery.

The descending aorta

bronchial arteries

oesophageal arteries

intercostal arteries.

THE ABDOMINAL AORTA
phrenic arteries

coelic artery $\left\{\begin{array}{l}\text{gastric arteries} \\ \text{splenic artery} \\ \text{hepatic artery}\end{array}\right.$

superior mesenteric artery

renal arteries

ovarian or testicular arteries

inferior mesenteric artery

right and left common iliac arteries $\left\{\begin{array}{l}\text{external iliac artery} \\ \text{internal iliac artery.}\end{array}\right.$

ARTERIES SUPPLYING THE HEAD AND NECK

The common carotid artery

External carotid artery: thyroid artery
    lingual artery
    facial artery
    occipital artery
    temporal artery
    maxillary artery.

Internal carotid artery: ophthalmic artery
    anterior cerebral artery
    middle cerebral artery
    posterior communicating artery.

The circulus arteriosus (circle of Willis):
    anterior cerebral arteries
    middle cerebral arteries
    posterior communicating arteries
    posterior cerebral arteries, which are branches
      from the basilar arteries.

VENOUS RETURN FROM THE HEAD AND NECK

Superficial veins

thyroid vein
facial vein $\left.\right\}$ empty into the external jugular vein.
occipital vein

Deep sinuses

superior saggital sinus
inferior sagittal sinus $\left.\right\}$ empty into the internal jugular vein.
straight sinus
transverse sinuses

NB There is only one superior sagittal sinus,
one inferior saggital sinus and one straight sinus.

ARTERIES SUPPLYING THE UPPER EXTREMITY

subclavian artery
axillary artery
brachial artery
radial artery
ulnar artery
deep and superficial palmar arch
digital arteries.

VENOUS RETURN FROM THE UPPER EXTREMITY

Deep veins

digital veins
palmar venous arch
radial vein
ulnar vein
brachial vein
axillary vein
subclavian vein
brachio-cephalic vein
superior vena cava.

Superficial veins

digital veins
palmar venous arch
cephalic vein
basilic vein
median cubital vein
median vein.

ARTERIES SUPPLYING THE LOWER EXTREMITY

external iliac artery
femoral artery
popliteal artery
anterior tibial artery
posterior tibial artery
plantar arch
digital arteries

VENOUS RETURN FROM THE LOWER EXTREMITY

Deep veins

digital veins
plantar veins
anterior tibial veins
posterior tibial veins
popliteal vein
femoral vein
external iliac vein
common iliac vein
inferior vena cava.

Superficial veins

digital veins
long saphenous vein
short saphenous vein.

8. The Lymphatic System

The lymphatic system communicates with the blood circulatory system and is closely associated with it.

As described previously the tissue fluid is derived from the blood plasma. A certain amount of this fluid and waste products from the cells is returned to the venous capillaries, but within the tissue spaces fine capillary vesels known as *lymphatic capillaries* begin which help to drain the waste products and water from the interstitial spaces.

Figure 151

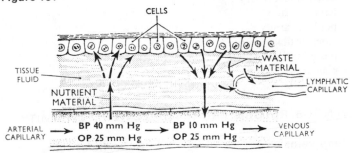

Scheme showing the flow of waste material into a lymphatic capillary.

As the pressure in the tissue spaces rises the fluid which they contain passes into the minute lymphatic capillaries, in the same way as into the venous capillaries, although the walls of the lymphatic capillaries have greater permeability than those of the blood capillaries.

The fluid within the lymphatic capillaries and vessels is known as *lymph*.

The composition of lymph is very like that of the blood plasma but the dissolved substances are in different concentrations. Lymph also contains materials which may be damaging to the body. Because of the greater permeability of the lymph capillaries substances of larger size can enter these vessels and be removed from the interstitial spaces, for example, if infection is present and phagocytosis has occurred the neutrophils and monocytes with their ingested micro-organisms are drained away in the lymphatic capillaries and vessels.

The lymphatic system consists of:
lymphatic capillaries
lymphatic vessels
the thoracic duct
the right lymphatic duct
lymph nodes.

Figure 152

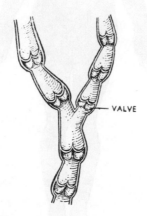

Interior of a lymphatic vessel.

Structure of lymphatic vessels

1. An outer coat consisting of *fibrous tissue* which acts as a protective covering.

2. A middle coat of *muscular and elastic tissue*.

3. An inner lining composed of a single layer of *endothelial cells*.

4. Lymphatic vessels contain many *valves* which give them a knotted or beaded appearance. The valves prevent the backward flow of the lymph.

The *lymphatic capillaries*, which are composed of very fine *connective tissue* and a single layer of *endothelial cells*, commence in the tissue spaces of the body as minute *blind end tubes*. These minute capillaries join with one another to form the *lymphatic vessels* until eventually *two main lymphatic ducts* are formed.

Figure 153

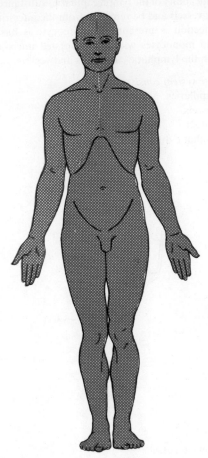

Schematic diagram showing the drainage of lymph. Red indicates lymph collected by the thoracic duct. Blue indicates lymph collected by the right lymphatic duct.

THE LYMPHATIC DUCTS

There are two lymphatic ducts which return lymph collected from the whole body and return it to the blood. Their structure is the same as that of the smaller lymphatic vessels.

The thoracic duct

This duct begins at the *cysterna chyli* which is a sac-like dilatation on the lymphatic pathway situated in front of the bodies of the first and second lumbar vertebrae to the right of the abdominal aorta.

The thoracic duct is the largest lymphatic vessel in the body and contains several valves. It is approximately 40 cm (16 inches) in length.

It extends from the lower border of the body of the 12th thoracic vertebra to the root of the neck and opens into the *left subclavian vein*.

This duct conveys all lymph from the lower limbs, pelvic cavity, abdominal cavity, *left* side of the chest, *left* side of the head, neck and *left* arm to the left subclavian vein.

The right lymphatic duct

The right lymphatic duct is about 1 cm (less than $\frac{1}{2}$ an inch) in length. It lies in the root of the neck and terminates by emptying its contents into the *right subclavian vein*. The right lymphatic duct receives all the lymph which has drained from the *right* side of the head and neck, the *right* arm and the *right* side of the chest.

LYMPH NODES

All the small and medium sized lymphatic vessels open into *lymph nodes* which are situated in strategic positions throughout the body. The lymph drains through at least one node before returning to the blood. These nodes vary considerably in size. Some are as small as a pin head and the largest are about the size of an almond.

Structure of lymph node

Lymph nodes have a *surrounding capsule of fibrous tissue* which dips down into the node substance forming partitions known as *trabeculae*.

Figure 154

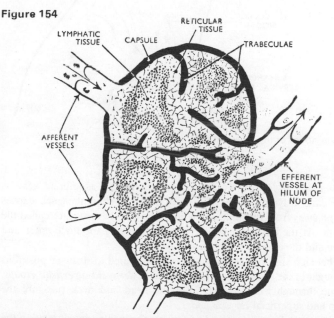

Diagrammatic illustration of a lymph node, showing the direction of flow of lymph.

The main substance of the node consists of *reticular and lymphatic tissue* containing many *lymphocytes*.

Figure 155

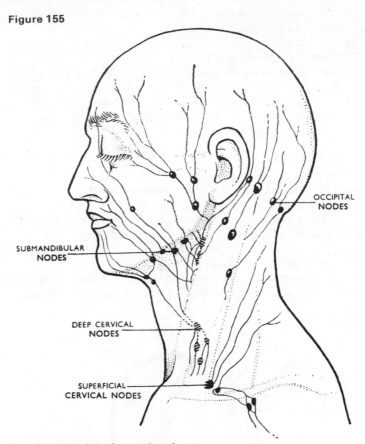

SUBMANDIBULAR
NODES

OCCIPITAL
NODES

DEEP CERVICAL
NODES

SUPERFICIAL
CERVICAL NODES

Lymph nodes of the face and neck.

As many as four or five lymph vessels, known as afferent vessels, may enter a lymph node while only one, the efferent vessel, carries lymph away from the node. Each node has a concave surface called the hilum. At the hilum the blood vessels supplying the node enter and leave and the efferent lymph vessel leaves.

There is a vast number of lymph nodes situated in strategic positions throughout the body. They are arranged in *deep and superficial groups*. Those through which lymph from the head and neck pass are the deep and superficial *cervical nodes*.

The lymph from the *upper limbs* passes through nodes situated in the elbow region then through the deep and superficial *axillary nodes*.

Lymph from the *thoracic cavity* is drained through many nodes, the more important groups are the *tracheobronchial* and the *intercostal nodes*.

Lymph from the *pelvic and abdominal cavities* is drained through many lymph nodes which are situated either in front of, or beside the aorta. These are the *pre-aortic* and the *lateral aortic nodes*.

The lymph from the *lower limbs* is drained through deep and superficial nodes behind the knee and in the groin which are known as the *popliteal* and the deep and superficial *inguinal nodes*.

Figure 156

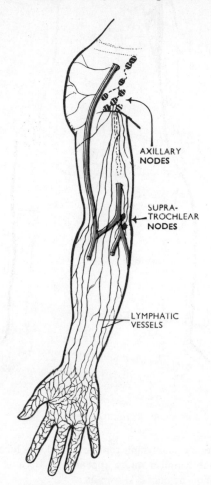

AXILLARY
NODES

SUPRA-
TROCHLEAR
NODES

LYMPHATIC
VESSELS

Diagram showing positions of lymph nodes of the arm.

The lacteals

The lacteals are the lymph capillaries which drain lymph from the small intestine. About 60 per cent of the fat absorbed from the small

Figure 157

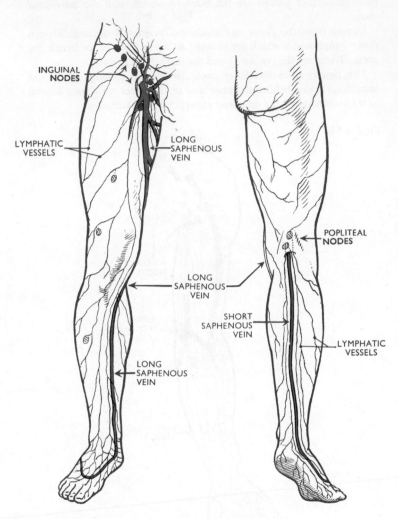

Diagram showing positions of lymph nodes of the leg.

intestine passes into the lymph capillaries and this high concentration of fat gives the lymph a milky appearance. Because of this lymph entering the thoracic duct is known as *chyle*.

Functions of the lymph nodes

1. Lymph is filtered as it passes through the lymph nodes leaving behind any particles which would not normally be found in serum.

2. The lymphoid tissue in the nodes breaks down material which has

been filtered off, for example, micro-organisms, phagocytes, tumour cells and cells which have been damaged by inflammation. The nodes are not always successful in destroying this type of material but, on the whole, they provide an effective barrier against the spread of noxious particulate material.

3. Lymphocytes develop from the reticular and lymphoid tissue in the nodes and pass to the blood via the lymph vessels and ducts.

4. Antibodies and antitoxins are formed by the cells in the lymph nodes.

Figure 158

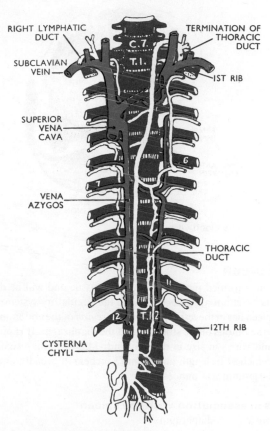

Formation and position of the thoracic duct.

LYMPHATIC TISSUE

Lymphatic tissue is found in a number of situations in the body in addition to the lymph nodes.

The tonsils	between the mouth and the oral pharynx.
The adenoids	on the wall of the nasopharynx.
The solitary lymphatic nodules	in the small intestine.
The aggregated lymphatic nodules (Peyers patches)	in the small intestine.
The vermiform appendix	associated with the large intestine.

These nodules are discussed in the digestive and respiratory systems.

Figure 159

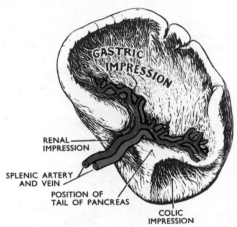

The spleen and its blood vessels.

The Spleen

The spleen is formed partly by lymphatic tissue and will be described here as its functions are associated with the circulatory system.

The spleen lies in the *left hypochondriac region* of the abdominal cavity between the fundus of the stomach and the diaphragm. It is purplish in colour and varies in size in different individuals, but is usually about 12 cm (5 inches) in length and 7 cm (3 inches) in breadth and weighs about 200 grammes (7 ounces).

Organs in association with the spleen

Superiorly	diaphragm
Inferiorly	left colic flexure of the large intestine
Anteriorly	fundus of the stomach
Posteriorly	left kidney and supra-renal gland
Medially	pancreas
Laterally	ninth, tenth and eleventh ribs and the intercostal muscles.

Figure 160

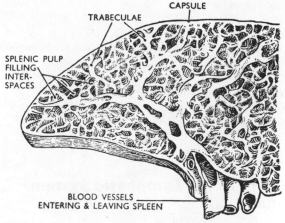

Scheme of the interior of the spleen.

STRUCTURE OF THE SPLEEN

The spleen is slightly ovoid in shape and is covered anteriorly by
peritoneum. Under the peritoneum there is a *fibro-elastic* capsule which
surrounds the gland. Fibrous tissue spreads into the organ substance to form *trabeculae*. The substance of the organ is known as the
splenic pulp. It is composed of lymphoid tissue and a large number of
blood capillaries.

The lower medial border of the organ is concave and is known as
the *hilum*. The vessels entering and leaving at the hilum are:

the splenic artery
the splenic vein
lymphatic vessels
nerves

FUNCTIONS OF THE SPLEEN

1. Destruction of worn-out erythrocytes

As described previously erythrocytes are destroyed in the spleen and
the breakdown products, bilirubin, biliverdin and iron are passed to the
liver via the splenic and portal veins.

2. A reservoir for blood

The fact that the spleen acted as a reservoir for blood was discovered
by experiments on animals. It was found that in certain conditions such
as a rise in temperature, strenuous exercise and emotional excitement,
there was an increase in blood pressure and an increase in the number of
circulating red blood corpuscles. This occurred due to strong
contractions of the spleen and the pushing out of extra blood into the
circulation.

Under normal circumstances the spleen contracts rhythmically about two or three times per minute.

3. The formation of lymphocytes
Due to the presence of lymphatic tissue the spleen produces some of the lymphocytes.

4. Formation of antibodies and antitoxins
Due to the presence of lymphoid reticulo-endothelial tissue the spleen forms antibodies and antitoxins.

Summary of the Lymphatic System
STRUCTURE
The lymphatic system consists of:
lymphatic capillaries
lymphatic vessels
lymphatic nodes
cisterna chyli
the thoracic duct
the right lymphatic duct.

Lymphatic capillaries
1. Blind end tubes starting in the periphery of the body and within the organs.
2. They are composed of a single layer of fine connective tissue and endothelial cells.
3. The lymphatic capillaries from the small intestine are termed lacteals and contain absorbed fat. The content of the lacteals is called chyle.

Lymphatic vessels
1. Have the same structure as veins with valves.
2. Those entering lymph nodes are called afferent vessels and those leaving are known as efferent vessels.

The cisterna chyli
This is a sac-like structure lying in front of the first two lumbar vertebrae. Receives all the lymph from the legs, pelvic cavity and abdominal cavity.

The thoracic duct
Conveys lymph from the legs, pelvic cavity, abdominal cavity, left side of the chest, left side of the head and neck and left arm to the left subclavian vein.

The right lymphatic duct

A very short duct which conveys lymph from the right side of the chest, head and neck and right arm to the right subclavian vein.

Lymph nodes

These are composed of lymphatic tissue enclosed within a fibrous capsule.

FUNCTIONS

1. Filtration of lymph.
2. Liquefaction and destruction of damaging substances.
3. Formation of antibodies and antitoxins.
4. Formation of lymphocytes.

Lymphatic nodes are placed in strategic positions throughout the body and lie in deep and superficial groups. Some of the important nodes are:

deep and superficial cervical nodes
deep and superficial axillary nodes
intercostal nodes
tracheobronchial nodes
pre-aortic nodes
lateral aortic nodes
deep and superficial inguinal nodes
popliteal nodes.

LYMPHATIC TISSUE IN OTHER PARTS OF THE BODY

The tonsils
The adenoids
The solitary lymphatic nodules
The aggregated lymphatic nodules
The vermiform appendix.

THE SPLEEN

The spleen lies in the left hypochondriac region of the abdominal cavity. It is surrounded by a fibro-elastic capsule. Its structure is mainly of splenic pulp. It is supplied by the splenic artery and drained by the splenic vein which joins the superior mesenteric vein to form the portal vein.

Functions of the spleen

Destruction of worn-out red blood corpuscles
Reservoir for blood
Formation of lymphocytes
Formation of antibodies and antitoxins.

9. The Respiratory System

The respiratory system is composed of various organs the structure of which ensure a clear pathway for air to enter and leave the lungs and the mechanism for the exchange of air between the lungs and the atmosphere.

The Approximate Composition of Air

Inspired Air		Expired Air	
	Per cent		*Per cent*
Oxygen	21	Oxygen	17
Nitrogen	78	Nitrogen	78
Carbon Dioxide	0·04	Carbon Dioxide	4·04
Inert gases, e.g. Helium		Inert gases, e.g. Helium	
Argon	1	Argon	1
Water vapour	Variable	Water Vapour	Saturated
Temperature	Variable	Temperature	Body temperature

The body requires a constant supply of oxygen from the air, and a means of disposing of carbon dioxide which is produced as a waste product of cell metabolism. The blood provides the transport system for these gases between the lungs and the cells of the body.

The exchange of gases between the blood and the lungs is called *external respiration* and that between the blood and the cells, *internal respiration*.

The organs of the respiratory system are:
   the nose
   the pharynx
   the larynx
   the trachea
   two bronchi
   the bronchioles and smaller air passages
   the two lungs and their coverings—the pleura.

Figure 161

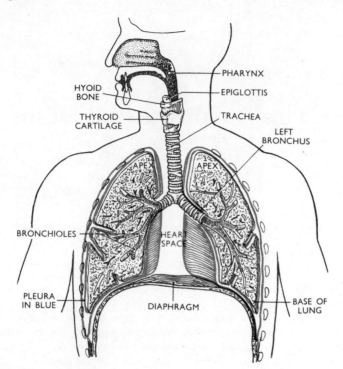

General view of respiratory organs.

Nose and Nasal Cavity

The nasal cavity is the first of the respiratory organs and consists of a large irregular cavity divided by a *septum*.

The roof is formed by the base of the skull.

The floor is formed by the roof of the mouth, which is composed of the maxillary bones, the palatine bones and the soft palate.

The medial wall is formed by the *septum* which is composed of the perpendicular plate of the ethmoid bone superiorly, and the vomer inferiorly. Anteriorly the septum is formed by cartilage.

The anterior wall is formed by the nasal bones and cartilage covered with skin.

The lateral walls are formed by the superior maxillary bones and the inferior, middle and superior conchae.

The posterior wall is formed by the posterior wall of the pharynx.

Figure 162

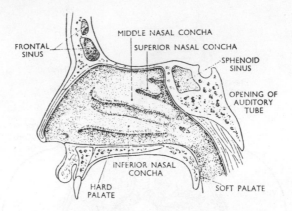

Interior of the nose showing conchae and some of the sinuses of the skull.

The anterior nares form the anterior opening into the nasal cavity.

The posterior nares form the posterior opening from the nasal cavity into the pharynx.

The nose is lined with *ciliated epithelium,* the area of which is increased by the presence of the conchae which project into the cavity. The presence of ciliated epithelium ensures that the air is filtered, moistened and warmed during its passage through the nose.

Figure 163

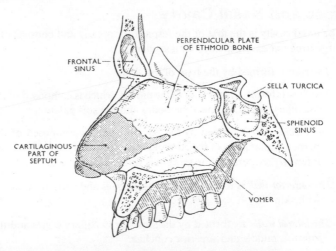

Diagram showing structures forming the septum of the nose.

Figure 164

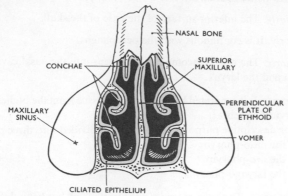

Interior of the nose showing extent of mucous membrane in yellow.

Openings into the nasal cavity
1. The anterior nares through which air passes from the exterior.
2. The posterior nares through which air passes into the naso-pharynx.
3. The openings into the maxillary sinuses in the lateral walls.
4. The openings into the frontal and sphenoidal sinuses in the roof of the nasal cavity.
5. The openings into the ethmoidal sinuses in the upper lateral walls between the superior and middle conchae.

The Pharynx
The pharynx is a cone-shaped tube part of which is common to the respiratory system and the alimentary canal. It is approximately 12 to 14 cm (5 inches) in length extending from the inferior surface of the base of the skull to the level of the sixth cervical vertebra. It is the shape of an inverted cone being wider at its upper end.

Figure 165

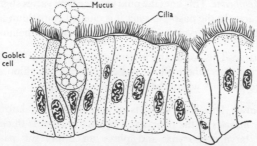

Illustration of ciliated columnar epithelium with goblet cells.

Organs in association with the pharynx

Superiorly. The inferior surface of the base of the skull.

Inferiorly. It is continuous with the oesophagus.

Anteriorly. The wall is incomplete and opens into the nasal cavity, the mouth and the larynx.

Posteriorly. It is separated by loose areolar tissue from the bodies of the six cervical vertebrae.

For descriptive purposes the pharynx is divided into three parts:
the naso-pharynx
the oro-pharynx
the laryngo-pharynx.

Naso-pharynx. The naso-pharynx is the part which lies behind the nose above the level of the soft palate. On its lateral walls are the two openings of the *auditory tubes* which lead from the naso-pharynx to the middle ear. On the posterior wall there are the *adenoids* (pharyngeal tonsil) which consist of lymphoid tissue. The adenoids are most prominent in children up to approximately the age of 7 years and thereafter begin to atrophy.

Oro-pharynx. The oro-pharynx is that part of the pharnyx behind the mouth extending from below the level of the soft palate to the level of the second cervical vertebra. The lateral walls of the oro-pharynx blend with the folds of the soft palate. Between these folds called the *pillars of the fauces* lie the *oral tonsils* which are collections of lymphoid tissue. The tonsils guard the opening into both the alimentary tract and the respiratory tract.

Laryngo-pharynx. The laryngo-pharynx extends from the oro-pharynx above and continues as the oesophagus below, i.e., from the level of the second to the sixth cervical vertebra.

STRUCTURE OF THE PHARYNX
The pharynx is composed of three layers of tissue.

Mucous Membrane. This lines the pharynx, the type varies slightly in the different parts. In the naso-pharynx it is ciliated epithelium and in the oro- and laryngo-pharynx it is stratified squamous epithelium.

Fibrous Tissue. This forms the intermediate layer. It is thicker at the naso-pharynx where there is little muscle and becomes thinner towards the laryngo-pharynx where the muscle layer is thicker.

Muscle Layer. This layer consists of several muscles known as the *constrictor muscles* of the pharynx. The muscle layer is thickest in the oro- and laryngo-pharynx and plays an important part in the mechanism of swallowing (deglutition).

Figure 166

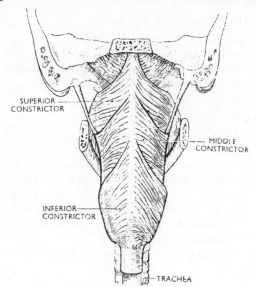

Illustration of constrictor muscles of the pharynx.

The blood supply to the pharynx is through several arteries which are branches of the facial artery.

The venous return is into the facial and internal jugular veins.

The nerve supply is from the glossopharyngeal nerve and its plexus.

The Larynx

The larynx or 'voice box' extends from the root of the tongue and the hyoid bone to the trachea. It lies opposite the third, fourth, fifth and sixth cervical vertebrae but is slightly higher in the female than in the male. Until puberty there is little difference in the size of the larynx in the different sexes, but after puberty there is considerable enlargement in the male as compared to the female, which explains the prominence of the 'Adam's apple' in the male In the adult female the larynx is approximately 3·5 cm (1¼ inches) in length and in the male 4·5 cm (1¾ inches).

Organs in association with the larynx

Superiorly. The hyoid bone and the root of the tongue.

Inferiorly. It is continued as the trachea.

Anteriorly. The muscles attached to the hyoid bone and the muscles of the neck.

Posteriorly. The laryngo-pharynx and cervical vertebrae.

Laterally. The lobes of the thyroid gland.

STRUCTURE OF THE LARYNX

The larynx is composed of several irregularly-shaped cartilages attached to each other by ligaments and membranes:

> the thyroid cartilage
> the cricoid cartilage
> the arytenoid cartilages
> the epiglottis.

The Thyroid Cartilage. This is the most prominent and consists of two flat pieces of cartilage known as the *laminae* which are fused together anteriorly to form the *laryngeal prominence* (Adam's apple). Immediately above the laryngeal prominence the laminae dip down to form a V shaped notch known as the *thyroid notch.* The thyroid cartilage is incomplete posteriorly and the laminae diverge to form two slender processes known as the *superior and inferior horns.*

Figure 167

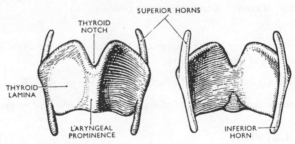

Diagram of thyroid cartilage

The upper part of the thyroid cartilage is lined with stratified squamous epithelium and the lower part with ciliated epithelium. There are many muscles attached to its outer surface.

Figure 168

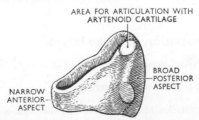

The cricoid cartilage.

The Cricoid Cartilage. This lies below the thyroid cartilage. It is shaped like a signet ring completely encircling the larynx with the narrow part in front and the broad part behind. It is lined with ciliated epithelium and there are muscles and ligaments attached to its outer surface.

The Arytenoid Cartilages. These are two roughly pyramidal shaped cartilages situated on top of the broad part of the cricoid cartilage forming part of the posterior wall of the larynx. They give attachment to muscles and are lined with ciliated epithelium.

The Epiglottis. This is a leaf-shaped cartilage attached to the inner surface of the anterior wall of the thyroid cartilage immediately below the thyroid notch. It rises obliquely upwards behind the tongue and the body of the hyoid bone. It is covered with stratified squamous epithelium.

Figure 169

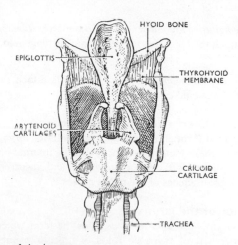

Posterior view of the larynx.

Ligaments and membranes of the larynx

The hyoid bone and the cartilages which form the larynx are attached to one another by ligaments and membranes.

The Thyrohyoid Membrane. This is a broad flat membrane composed of fibro-elastic tissue attached to the lower border of the hyoid bone above and to the upper border of the thyroid laminae and the superior horns of the thyroid cartilage below.

The Cricothyroid Ligament and Cricovocal Membrane. These are composed mainly of yellow elastic tissue and extend from the lower border of the laminae and inferior horns of the thyroid cartilage above to the upper border of the cricoid cartilage below.

Figure 170

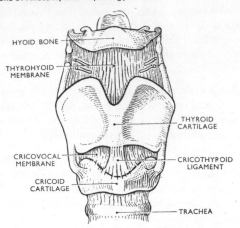

Diagram showing the membranes and ligaments of the larynx.

The Thyroepiglottic Ligament and Hyoepiglottic Ligaments. These ligaments attach the epiglottis to the thyroid cartilage and hyoid bone respectively.

The Cricoarytenoid Ligaments. These ligaments attach the arytenoid cartilages to the cricoid cartilage.

The blood and nerve supply to the larynx

Blood is supplied through the superior and inferior laryngeal arteries and drained by the thyroid vein which joins the internal jugular vein.

The nerve supply is from the laryngeal and recurrent laryngeal nerves which are branches of the vagus nerves.

The interior of the larynx

Within the larynx lie the *vocal cords*, thus the term 'voice box'. The vocal cords consist of two pale folds of mucous membrane which extend from the inner wall of the thyroid prominence anteriorly to be attached to the arytenoid cartilages posteriorly.

When the muscles which surround the arytenoid cartilages contract they pull the vocal cords together narrowing the gap between them, thus the *chink of the glottis* is formed. If air is forced through this chink it causes vibration of the cords and sound is produced. When the muscles surrounding the arytenoid cartilages relax the cartilages move apart and thus the cords are separated and no sound is produced.

Sound has the properties of *pitch, loudness* and *quality.*

The pitch of the voice depends upon the *length* and *tightness* of the cords. In the male the vocal cords are longer than in the female, 2·3 cm as against 1·7 cm. Thus the male voice has a deeper pitch than the female.

Figure 171

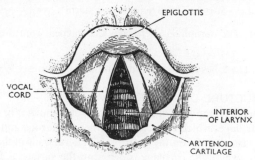

Interior of larynx showing position of vocal cords.

The loudness of the voice depends upon the *force* with which the cords vibrate. The greater the force of expired air the more vibration of the cords, thus the louder the sound.

The quality and resonance of the voice depend largely upon the shape of the mouth, the position of the tongue, the lips, the facial muscles and on the air sinuses in the bones of the face.

Figure 172

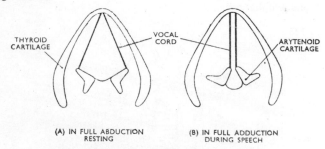

Diagram showing different positions of the vocal cords.

The Trachea

The trachea or wind pipe is a continuation of the larynx and extends to approximately the level of the fifth thoracic vertebra where it divides or *bifurcates* into the right and left bronchi. It is approximately 10 to 11 cm (4½ inches) in length. It lies mainly in the median plane in front of oesaphagus.

Organs in association with the trachea

Superiorly. The larynx.

Inferiorly. The right and left bronchi.

Anteriorly. Upper part—the isthmus of the thyroid gland. Lower part—the arch of the aorta and the manubrium sterni.

Posteriorly. The oseophagus separates the trachea from the vertebral column.

Laterally. The lobes of the thyroid gland and the lungs.

THE STRUCTURE OF THE TRACHEA

The trachea is composed of from 16 to 20 incomplete, or C shaped hyaline cartilages which are joined together by fibrous and muscular tissue. The cartilage forms the anterior and lateral walls of the tube and posteriorly, where the cartilage is deficient, the surface is flat and is composed of fibrous and elastic tissue with involuntary muscle fibres. The C shaped cartilages are enclosed by fibrous and elastic tissue. These blend to form a stout membrane which connects the rings to one another.

Coverings of the trachea

Outer. The outer surface of the trachea is composed of fibrous and elastic tissue which encloses the cartilages.

Middle. The middle layer is composed of a thin layer of fibrous and elastic tissue lined with areolar tissue through which blood vessels, nerves and lymphatics pass.

Figure 173

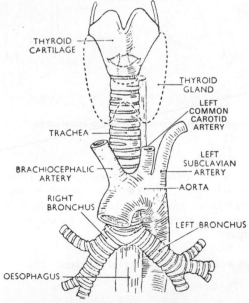

Diagram of organs in association with the trachea.

Inner. The inner lining is of ciliated epithelium containing goblet cells which secrete mucus.

The blood supply to the trachea is mainly by the inferior thyroid arteries.

The venous return is through the inferior thyroid veins into the brachio-cephalic vein.

The nerve supply is by the vagus and recurrent laryngeal nerves.

The Bronchi

The two bronchi commence when the trachea bifurcates, that is, at about the level of the 5th thoracic vertebra.

The Right Bronchus. This is a wider shorter tube than the left bronchus and it lies in a more vertical position. It is approximately 2·5 cm (1 inch) in length. After entering the right lung at the hilum it divides into three branches one of which passes to each lobe. Each branch then subdivides into numerous smaller branches.

Figure 174

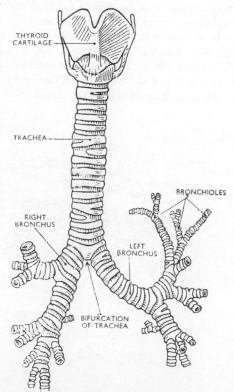

THYROID
CARTILAGE

TRACHEA

BRONCHIOLES

RIGHT
BRONCHUS

LEFT
BRONCHUS

BIFURCATION
OF TRACHEA

The trachea, bronchi and bronchioles.

The Left Bronchus. This is narrower and longer than the right, being about 5 cm (2 inches) in length. After entering the lung the left bronchus divides into two branches, one of which goes to each lobe. Each branch then subdivides into progressively smaller tubes within the lung substance.

THE STRUCTURE OF THE BRONCHI

The bronchi are composed of the same tissues as the trachea. As the bronchi became smaller, by subdividing, the cartilages became less well defined and more irregular in shape. The bronchi are lined with ciliated epithelium.

The Bronchioles

This is the name given to the smaller branches of the bronchi which have a diameter of about 1 mm.

The bronchioles contain no cartilage in their walls, but are initially composed of muscle tissue, fibrous tissue and elastic tissue with an inner lining of ciliated columnar epithelium. Gradually as the tubes become smaller the fibrous and muscle tissue disappears, and the columnar epithelium changes to a single layer of flattened epithelial cells. The minute bronchioles, known as the *terminal bronchioles* branch to form respiratory bronchioles which branch still farther to form *alveolar ducts*. The alveolar ducts then lead into minute sac-like structures known as the *alveoli*. It is here that the interchange of gases takes place between the air in the alveoli and the blood in the capillaries, which are also composed of a single layer of flattened epithelial cells.

Figure 175

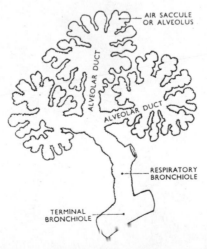

Diagram illustrating the terminal bronchioles and alveoli.

The Lungs

The lungs are two in number and lie in the thoracic cavity separated from each other by the heart and great blood vessels. They extend from the root of the neck above to the diaphragm below. They are roughly conical in shape with the *apex* above and the *base* below.

Organs in association with the lungs

Superiorly. The structures at the root of the neck.

Inferiorly. The diaphragm.

Anteriorly. The ribs, costal cartilages and intercostal muscles. At the apex the subclavian artery and vein.

Posteriorly. The ribs and intercostal muscles, and transverse processes of the thoracic vertebrae.

Medially. The heart, aorta, pulmonary artery, venae cavae, oesophagus, trachea, bronchi and the bodies of the thoracic vertebrae.

Laterally. The ribs and intercostal muscles.

The space between the medial borders of the lungs in which the heart and great blood vessels and the oesophagus lie is known as the *mediastinum*.

Figure 176

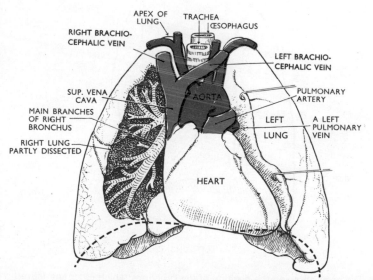

Diagram showing the organs in association with the lungs.

Figure 177

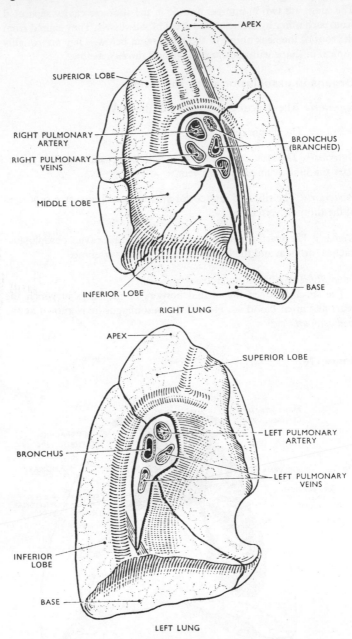

RIGHT LUNG

LEFT LUNG

Diagrams showing the position of the structures which enter and leave each lung at the hilum.

THE STRUCTURE OF THE LUNGS

The right lung is divided into three distinct lobes:
   superior
   middle
   inferior.

The left lung is divided into two lobes:
   superior
   inferior.

Each lung is described as having a costal and a medial surface.

The costal surface is smooth and follows the shape of the chest wall.

The medial surface is concave in appearance. The concavity is greater in the left lung than the right due to the presence of the heart which is situated in the mediastinum slightly to the left of the midline. Above and a little behind the concavity is a roughly triangular shaped depression known as the *hilum of the lung*. The structures forming the *root of the lung* enter and leave at the *hilum*.

The roots of the lungs are at the level of the fifth, sixth and seventh thoracic vertebrae.

Structures entering and leaving a lung at the root

   A bronchus.
   A pulmonary artery.
   A bronchial artery.
   Branches of the parasympathetic and sympathetic nerves.
   Two pulmonary veins.
   The bronchial veins.
   Lymphatic vessels.

The pleura

Each lung is invested by a serous membrane—*the pleura*.

The pleura is composed of flattened epithelial cells which lie on a basement membrane. There are two layers of membrane:
   parietal pleura
   visceral pleura.

The Parietal Pleura. This lines the ribs, sternum, costal cartilages and the internal intercostal muscles and covers the superior surface of the diaphragm.

The Visceral Pleura. This is firmly attached to the lung itself, completely covering its surfaces and passing into the fissures which divide the lung into its lobes. At the root of each lung it is reflected and becomes the parietal pleura.

In health the parietal and visceral pleura are in *close contact* separated only by a thin film of serous fluid. There is a *potential space* between the layers which is termed the *pleural cavity*. The serous fluid which separates the two layers of pleura and prevents friction between them is secreted by the epithelial cells of the membrane.

The interior of the lungs

The lungs are composed of the bronchi, bronchioles, alveolar ducts and alveoli of which there are many thousands. The respiratory bronchioles, alveolar ducts and alveoli constitute the *lobules* of the lungs. These lobules are separated from one another by the interlobular tissue which is composed of *areolar and elastic tissue.*

The pulmonary artery divides into a right and left branch, one of which conveys deoxygenated blood to each lung. On entering the lung the pulmonary artery divides into many branches which accompany the bronchi and bronchioles. They subsequently end in a dense capillary network in the walls of the alveoli. The walls of the alveoli and of the capillaries consist of only one layer of flattened epithelial cells. Therefore, the exchange of gases between the air in the alveoli and the blood in the capillaries takes place across these two very fine membranes. The pulmonary capillaries join up, eventually becoming two main pulmonary veins, which leave the hilum of each lung conveying oxygenated blood to the left atrium of the heart. The innumerable blood capillaries and blood vessels in the lungs are supported by *areolar and elastic tissue.*

Figure 178

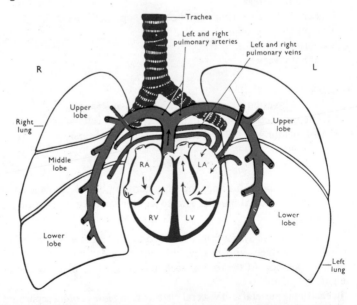

Schematic diagram showing the flow of blood between the heart and the lungs.

The Functions of the Respiratory Organs

The general function of the respiratory system is to ensure the intake of air into the lungs and the transmission of oxygen from the alveolar air to the blood, and the excretion of carbon dioxide from the blood and its passage through the respiratory tract to the atmospheric air.

Figure 179

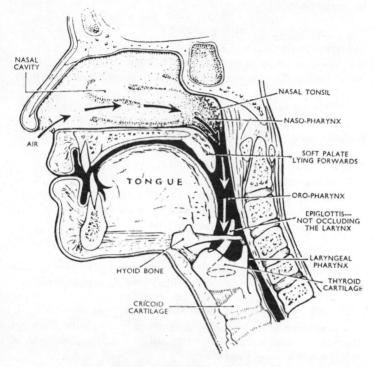

NASAL CAVITY

AIR

TONGUE

NASAL TONSIL

NASO-PHARYNX

SOFT PALATE LYING FORWARDS

ORO-PHARYNX

EPIGLOTTIS— NOT OCCLUDING THE LARYNX

LARYNGEAL PHARYNX

THYROID CARTILAGE

HYOID BONE

CRICOID CARTILAGE

Diagram showing the pathway of air from the nose to the larynx.

FUNCTION OF THE NOSE

The nose is the first of the respiratory passages through which the incoming air passes. The function of the nose is to ensure that the air is *warmed, moistened* and *filtered.*

The nose is lined with ciliated mucous membrane which is richly supplied with blood and its surface area is increased by the conchae.

The air is warmed as it passes over the surface of the nose; it is moistened by contact with the moist mucus; and it is filtered in the sense that particles of dust and other impurities in the air stick to the mucus.

The cilia of the mucous membrane waft the mucus towards the throat and it is swallowed.

THE FUNCTIONS OF THE PHARYNX

The pharynx is an organ involved both in the respiratory system and digestive system, therefore both air and food pass through it. By the same methods as in the nose the air is further warmed and moistened as it passes through the pharynx.

The auditory tubes pass between the naso-pharynx and the middle ear, and it is through these tubes that air passes to the middle ear. The presence of air in the middle ear at atmospheric pressure is essential for satisfactory hearing.

THE FUNCTIONS OF THE LARYNX

The larynx has three functions.

1. To ensure the passage of air from the pharynx above to the trachea below. Further filtering, moistening and warming of the air takes place in the larynx.

2. To ensure voice production due to the presence of the vocal cords.

3. To rise up so that the laryngeal opening is occluded by the epiglottis during swallowing or deglutition.

THE FUNCTIONS OF THE TRACHEA, BRONCHI AND BRONCHIOLES

The function of these organs is to ensure the passage of air from the larynx to the alveoli of the lungs. The air is warmed, moistened and filtered in its passage through these organs.

THE FUNCTIONS OF THE LUNGS

The function of the lungs is to allow a free exchange of gases to take place between the alveoli and the capillary network around them.

THE MECHANISM OF RESPIRATION

The mechanism of respiration is the process by which the lungs are expanded to take in air then contracted to expel it. The cycle of respiration consists of three phases:

inspiratory phase
expiratory phase
pause.

This cycle of events occurs about 16 times per minute. The expansion of the chest during inspiration occurs as a result of muscular activity which is partly voluntary and partly involuntary. The main muscles of respiration in normal quiet breathing are the *intercostal muscles* and the *diaphragm*.

The intercostal muscles

There are 11 pairs of intercostal muscles which occupy the spaces between the ribs. They are arranged in two layers called:

the external intercostal muscles
the internal intercostal muscles.

The External Intercostal Muscle Fibres. These extend in a *downwards and forwards* direction from the lower border of the rib above to the upper border of the rib below.

Figure 180

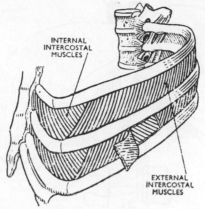

The internal and external intercostal muscles.

The Internal Intercostal Muscle Fibres. These extend in a *downwards and backwards* direction from the lower border of the rib above to the upper border of the rib below, crossing the external intercostal muscle fibres at right angles.

The first rib is fixed, therefore, when the intercostal muscles contract they *pull* all the other ribs towards the first rib. Because of the shape of the ribs they *move outwards* when they are *pulled upwards.* In this way the thoracic cavity is enlarged *antero-posteriorly and laterally.* The intercostal muscles are stimulated to contract by the *intercostal nerves.*

The diaphragm

The diaphragm, when *relaxed,* is a dome-shaped structure which separates the thoracic cavity from the abdominal cavity. It forms the floor of the thoracic cavity and the roof of the abdominal cavity and consists of a central tendon from which muscle fibres radiate to be attached to the vertebral column, the lower ribs and the sternum. When the muscle of the diaphragm is relaxed the central tendon is at the level of the 8th thoracic vertebra.

When the *diaphragm contracts,* its muscle fibres shorten and the central tendon is *pulled downwards,* enlarging the thoracic cavity in *length.* This increases the pressure in the abdominal and pelvic cavities. The diaphragm is supplied by the *phrenic nerves.*

It is important to appreciate that the intercostal muscles and the diaphragm contract *simultaneously* thus ensuring the enlargement of the thoracic cavity in all directions, that is from back to front, side to side and from top to bottom.

Figure 181

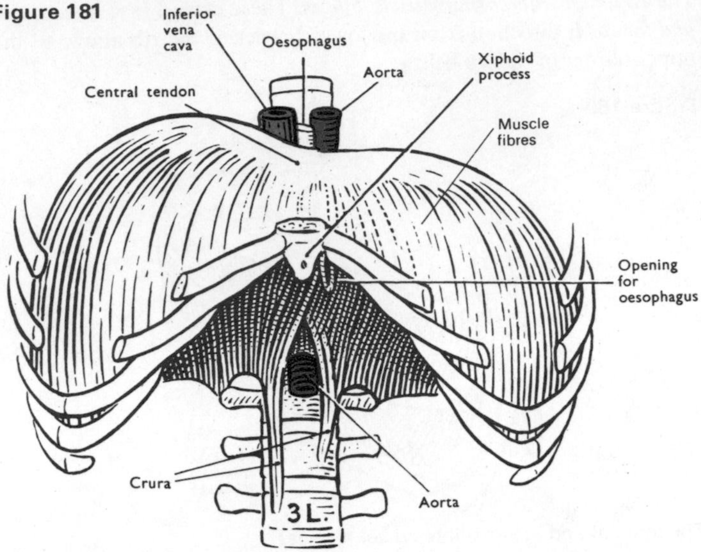

The diaphragm.

CHANGES IN THE LUNGS DUE TO CONTRACTION OF THE MUSCLES OF RESPIRATION

As described previously the visceral pleura is adherent to the lungs, the parietal pleura to the inner wall of the thorax and to the diaphragm and there is a potential space between these two layers of serous membrane called the pleural cavity.

When the capacity of the thoracic cavity is increased by simultaneous contraction of the intercostal muscles and the diaphragm, the parietal pleura moves with the walls of the thorax and the diaphragm. This reduces the pressure in the pleural cavity to a level considerably lower than atmospheric pressure and the visceral pleura tends to follow the parietal pleura. During this process the lungs are stretched and the pressure within the alveoli and in the air passages is reduced. This results in air being drawn into the lungs in an attempt to equalise the atmospheric and alveolar air pressures.

This is the process of *inspiration* which is described as *active* because it is the result of muscle contraction. When the diaphragm and intercostal muscles *relax* the ribs fall back into place, the diaphragm ascends, the lungs recoil and *expiration occurs*. This is a *passive* process. After expiration there is a short pause.

The interchange of gases within the lungs

Between the process of inspiration and expiration the interchange of gases takes place. This interchange occurs between the blood in the capillary network which surrounds the alveoli and the air in the alveoli of the lungs.

Figure 182

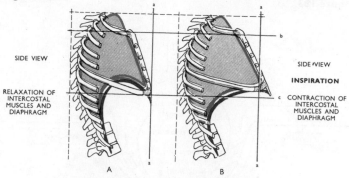

SIDE VIEW

RELAXATION OF
INTERCOSTAL
MUSCLES AND
DIAPHRAGM

SIDE VIEW

INSPIRATION

CONTRACTION OF
INTERCOSTAL
MUSCLES AND
DIAPHRAGM

A   B

(1) Outward movement of ribs as shown by line a–a
(2) Upward movement of ribs as shown by lines b and c
(3) Change in position of diaphragm due to its contraction
(4) The resulting increase in the size of the lungs

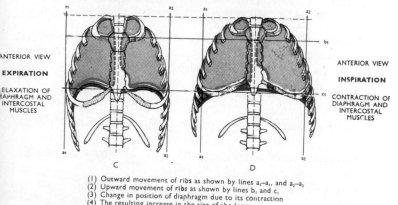

ANTERIOR VIEW

EXPIRATION

RELAXATION OF
DIAPHRAGM AND
INTERCOSTAL
MUSCLES

ANTERIOR VIEW

INSPIRATION

CONTRACTION OF
DIAPHRAGM AND
INTERCOSTAL
MUSCLES

C   D

(1) Outward movement of ribs as shown by lines a_1–a_1 and a_2–a_2
(2) Upward movement of ribs as shown by lines b_1 and c_1
(3) Change in position of diaphragm due to its contraction
(4) The resulting increase in the size of the lungs

Scheme showing changes in the size of the thoracic cavity due to
contraction of the muscles of respiration.

Elementary Physical Laws in relation to Gases. These state that:

1. the molecules of gases are always in motion
2. gases always tend to diffuse from a *higher partial pressure*
to a *lower partial pressure*, i.e., down the concentration gradient.
3. gases always exert pressure upon the walls of their container.

The atmospheric pressure at sea level is 760 millimetres of mercury
(mmHg). This pressure is exerted by the mixture of gases which make
up the air:

oxygen	21 per cent
nitrogen	78 per cent
carbon dioxide	0·04 per cent
rare gases	1 per cent
water vapour	variable.

Figure 183

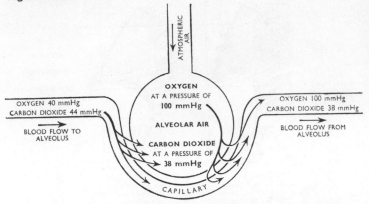

Schematic illustration of the interchange of gases between the alveoli of the lungs and the capillaries within the lungs.

Each one of these gases exerts a part of the total pressure depending upon its concentration in the mixture. The proportion of the total pressure provided by each gas is called its *partial pressure*.

Partial pressure of oxygen $(pO_2) = \dfrac{21}{100}$ of 760 = 160 mmHg

Partial pressure of nitrogen $(pN_2) = \dfrac{78}{100}$ of 760 = 592 mmHg

Partial pressure of carbon dioxide $(pCO_2) = \dfrac{\cdot 04}{100}$ of 760 = 0·3 mmHg

Partial pressure of rare gases $= \dfrac{1}{100}$ of 760 = 7·6 mmHg.

These figures vary slightly according to the amount of water vapour in the air.

During respiration the lungs and the respiratory passages are never empty of air. Instead there is a *tidal volume* of air (about 500 ml) which passes into and out of the lungs and air passage during each cycle or respiration in quiet breathing. The ebb and flow of air results in inspired air being mixed with the air in the lungs and the net result is that the pO_2 in the alveoli remains fairly constant at about 100 mmHg. Similarly the pCO_2 is about 38 mmHg.

In the deoxygenated blood in the capillaries of the pulmonary arteries surrounding the alveoli the pO_2 in solution is about 40 mmHg and the pCO_2 is about 44 mmHg.

As gases diffuse down the concentration gradient, oxygen diffuses from the alveoli to the blood and carbon dioxide from the blood to the alveoli. The blood in the capillaries moves slowly which allows sufficient time for the interchange of gases to take place and for the uptake of oxygen by the erythrocytes. By this mechanism carbon dioxide is excreted from the body and oxygen is absorbed and transported round the body in solution in the blood water and in combination with haemoglobin in the erythrocytes.

Figure 184

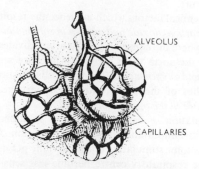

Diagram showing the capillary network surrounding the alveoli.

THE CONTROL OF RESPIRATION

The control of respiration is partly *chemical* and partly *nervous* but these are too closely linked to be described separately.

The respiratory centre is a collection of highly specialised cells situated in the medulla oblongata which are essential to normal

Figure 185

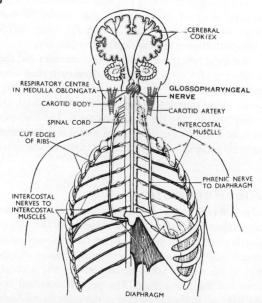

Scheme showing the nervous control of respiration.

respirations. The respiratory centre initiates nerve impulses which pass out from the brain in the *phrenic nerves* to the *diaphragm* and *intercostal nerves* to the *intercostal muscles*. These impulses stimulate the muscles to contract and increase the capacity of the thoracic cavity and so inspiration occurs.

The main chemical factors which influence the respiratory centre are *the amount of carbon dioxide and oxygen in solution in the blood*. In the carotid body, where the common carotid artery divides, and in the wall of the arch of the aorta there are cells which are sensitive to *carbon dioxide excess and oxygen lack*. When these cells are stimulated impulses pass in branches of the vagus and glossopharyngeal nerves to the *respiratory centre in the medulla oblongata*.

The accumulation of carbon dioxide in solution in the blood is therefore the most important stimulus to respiration. A lack of oxygen in the blood has some stimulating effect but if the pO_2 is greatly reduced the cells of the respiratory centre become less sensitive and do not function satisfactorily.

Approximately 4,000 to 6,000 millilitres (140 to 210 ounces) of blood flows through the lungs every minute and during that time. approximately 400 millilitres of oxygen are taken up by the blood. The amount of carbon dioxide excreted in expired air each minute is approximately 320 millilitres.

THE RATE AND DEPTH OF BREATHING

The rate and depth of breathing can vary considerably depending upon the health, physical activity and emotional state of the individual. The amount of air which enters and leaves the lungs varies mainly in association with changes in the depth of respiration.

In normal quiet breathing the intercostal muscles and the diaphragm are the only muscles involved, but in deep or forced breathing other muscles come into play, these are termed the *accessary muscles of respiration*. These muscles include the sterno-mastoid, the pectoralis major, the platysma, and the latissimus dorsi. The contraction of these muscles in addition to the diaphragm and intercostal muscles ensures the maximum increase in the capacity of the thoracic cavity.

Internal Respiration

Internal or cell respiration, it will be remembered, is the name given to the interchange of gases between the blood and the cells of the body.

Oxygen is carried from the lungs to the tissues dissolved in plasma and in chemical combination with haemoglobin, as *oxyhaemoglobin*. The exchange in the tissues takes place between the arterial end of the capillaries and the tissue fluid. The process involved is the same as that which occurs in the lungs, that is, *diffusion from a higher concentration of oxygen in the blood to a lower concentration in the tissue fluid*.

Figure 186

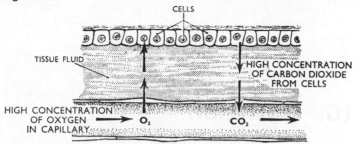

Schematic diagram of internal respiration.

Oxyhaemoglobin is an unstable compound which breaks up easily to liberate oxygen. As the cells of the body require a constant supply of oxygen the process of diffusion of oxygen from the blood across the capillary wall to the tissue fluid and then into the cells is continuous. The rate at which this process is carried on is increased in the presence of higher than usual concentration of carbon dioxide and this occurs when cells in a particular area are more than usually active. The higher pCO_2 assists the release of oxygen from oxyhaemoglobin. In this way cells which are particularly active receive an adequate oxygen supply.

Carbon dioxide is one of the waste products of carbohydrate and fat metabolism in the cells. The mechanism of transfer of carbon dioxide from the cells into the blood, at the *venous end of the capillary* is also by *diffusion*. Blood transports carbon dioxide by three different mechanisms:

1. some of the carbon dioxide is dissolved in the water of the blood plasma
2. some is transported in chemical combination with sodium in the form of sodium bicarbonate
3. the remainder is transported in combination with haemoglobin.

10. Nutrition

Before going on to discuss the subject of the digestive system it is necessary to have an understanding of the needs of the body in respect of diet.

The essentials of the diet include:

    carbohydrates
    proteins
    fats
    vitamins
    mineral salts
    water
    roughage.

If the cells of the body are to be able to function efficiently these nutritional necessities must be available in the *correct proportions.*

Carbohydrates are to be found in sugar, jam, cereals, bread, biscuits, potatoes, fruit and vegetables. They consist of carbon, hydrogen and oxygen, the hydrogen and oxygen being in the same proportion as in water.

Carbohydrates are classified according to the complexity of the chemical substances of which they are formed.

Monosaccharides. These are, chemically, the simplest form in which a carbohydrate can exist. They are made up of single units or molecules which, if they were broken down further, would cease to be monosaccharides. Carbohydrates are absorbed from the alimentary canal as monosaccharides and more complex carbohydrates are broken down to this form by digestion.

Glucose, fructose and galactose are examples of monosaccharides.

Disaccharides. These consist of two monosaccharide molecules chemically combined.

Sugars, such as, sucrose, maltose and lactose are disaccharides.

Polysaccharides. These consist of complex molecules made up of large numbers of monosaccharide molecules in chemical combination.

Starches, glycogen, cellulose and dextrins are examples of polysaccharides. Not all polysaccharides can be digested by human beings.

Functions of carbohydrates in the body

1. To provide energy and heat.
2. They act as a protein sparer. When there is an adequate supply of carbohydrate in the diet protein does not require to be used to provide energy and heat.
3. If excess carbohydrate is taken into the body it is deposited as fat in the fat depots.

Proteins or Nitrogenous Foods

Proteins consist of carbon, hydrogen, oxygen, nitrogen, sulphur and phosphorus. During digestion proteins are broken down to their simplest form, that is, amino acids, and it is as amino acids that they are absorbed.

Amino acids are divided into two categories, *essential* and *non-essential.*

Essential Amino Acids. These are the amino acids which must be eaten in the diet because they cannot be synthesised in the body.

Non-essential Amino Acids. These amino acids can be synthesised from other amino acids in the body.

First class protein is the name given to protein foods which contain all the essential amino acids in the correct proportions. They are derived almost entirely from animal sources and include:

meat	fish	eggs	chicken
milk	milk products	game	soya beans

Second class proteins are so called because they do not contain all the essential amino acids in the correct proportions. They are mainly of vegetable origin, examples are, peas, beans and lentils which are known as the pulses. A small proportion of protein is to be found in other vegetables.

Functions of protein in the body

1. To provide the amino acids required for the formation, growth and repair of body cells and for the formation of some of the secretions which they produce, e.g., hormones.
2. To provide the amino acids required for the formation of proteins which are present in the blood, i.e., serum albumen, serum globulin, fibrinogen and prothrombin.
3. To provide energy and heat, but only when there is an insufficiency of carbohydrate in the diet.

Fats

These foods consist of carbon, hydrogen and oxygen, but they differ from carbohydrates in that the hydrogen and oxygen are not in the same proportions as in water.

Fats are divided into two groups, animal and vegetable.

Animal Fat. This is found in milk, cheese, butter, eggs, meat, bacon and oily fish, for example, herring, cod and halibut.

Vegetable Fat. This is found in margarine where vitamins A and D are added and in nuts, for example, groundnuts and hazelnuts.

Functions of fat in the body

1. To produce energy and heat.
2. To support certain organs of the body, for example, the kidneys and the eyes.
3. To transport the fat soluble vitamins, which are A, D, E, and K.
4. Fat is necessary for the nerve sheaths, for the cholesterol in the bile, and the secretions of the sebaceous glands in the skin.

Vitamins

Vitamins are chemical compounds which are essential for health. They are found widely distributed in food. They are divided into two main groups:

fat soluble vitamins which are A, D, E, and K.
water soluble, which are B complex, C, and P.

FAT SOLUBLE VITAMINS

VITAMIN A

This vitamin is found in such foods as cream, egg yolk, fish oil, milk, cheese and butter. It can be formed in the body from certain carotenes of which the main dietary sources are green vegetables and carrots.

Vitamin A is only absorbed from the small intestine satisfactorily if fat absorption is normal.

Functions

1. It influences the nutrition of epithelial cells, thus tending to reduce the incidence and severity of micro-organism infection. Because of this, it is sometimes known as the *anti-infective vitamin.*
2. It is necessary for the regeneration of the visual purple in the retina of the eye which encourages rapid sight adaptation in the dark.
3. It is necessary to maintain the cornea of the eye in a healthy state.

VITAMIN D_3

Vitamin D_3 is sometimes termed the *antirachitic vitamin.* It is found mainly in animal fats such as eggs, butter, cheese, cod and halibut liver oils.

Man and animals can synthesise cholicalciferol (Vitamin D_3) by the action of the ultra-violet rays of the sun on a form of cholesterol in the skin known as 7-dehydrocholesterol.

Calciferol (Vitamin D_2) is formed in plants and is used widely in therapeutics.

Functions
This is the vitamin which regulates calcium and phosphorus metabolism and so is associated with the calcification of bones and teeth.

VITAMIN E OR TOCOPHEROL
The sources of this vitamin are peanuts, lettuce, egg yolk, wheat germ, whole cereal, milk and butter.

Functions
Lack of this vitamin in animals causes muscle wasting, and failure in reproduction, but there is no conclusive proof that the same holds good for human beings.

VITAMIN K
The sources of vitamin K are fish, liver, leafy green vegetables and fruit.

Bile salts must be present in the small intestine before it can be absorbed.

Functions
Vitamin K is necessary for the formation of prothrombin by the liver. Prothrombin, in turn, is necessary for the coagulation of blood.

WATER SOLUBLE VITAMINS

VITAMIN B COMPLEX
This comprises a group of water soluble vitamins which are more or less closely associated.

THIAMINE OR VITAMIN B_1
This vitamin is present in the germ of cereals, nuts, yeast, egg yolk, liver and legumes.

Functions
1. It is essential for the normal carbohydrate metabolism.
2. It stimulates appetite.
3. It helps to regulate the functioning of the nervous system.
4. It is associated with the control of water balance in the body.

RIBOFLAVIN OR VITAMIN B_2
This is found in yeast, leafy vegetables, milk, liver, eggs, kidney, cheese, roe.

Functions
1. It is concerned with the oxidation of all foods.
2. It is associated in some way with the physiology of vision.
3. It is necessary for the growth of all tissues in man and animals.

FOLIC ACID
This is found in liver, kidney, fresh deeply green vegetables and yeast.

It is synthesised by bacteria in the large intestine and absorbed in significant amounts.

Functions
It is associated with the development of the erythrocytes in the red bone marrow.

NICOTINIC ACID
This is found in liver, cheese, yeast, whole cereal, eggs, fish, peanuts and bemax.

Functions
1. It is necessary for the metabolism of carbohydrates.
2. It is necessary for the normal functioning of the gastro-intestinal tract.
3. It is required for the satisfactory functioning of the nervous system.

PYRIDOXINE OR VITAMIN B$_6$
This is found in egg yolk, peas, beans, soya bean, yeast, meat, liver.

Functions
It is believed to be necessary for satisfactory protein metabolism.

CYANOCOBALAMIN OR VITAMIN B$_{12}$
This is found in liver, meat, eggs, milk and fermented liquors.

Functions
It is necessary for the maturation of erythrocytes in the red bone marrow.

VITAMIN C AND VITAMIN P
These two vitamins are closely associated with each other, and their functions in the body are more or less similar.

They are found in fresh fruit, especially blackcurrants, oranges, lemons, also in rose-hips and green vegetables.

Functions
1. They are necessary for the maintainance of the strength of the walls of the blood capillaries.

2. They are necessary for healthy bones and teeth.
3. They are necessary for the formation of red blood cells.
4. They assist in some way with the production of antibodies.
5. They are associated with the synthesis of connective tissue.

Mineral Salts

Mineral salts (inorganic compounds) are necessary within the body for all body processes. They are usually required in small quantities.
They consist of soluble compounds of:

calcium	phosphorus	sodium
iron	iodine	potassium

CALCIUM

This is found in milk, cheese, eggs, green vegetables and some fish. The normal daily requirement for an adult is about 1 gramme. An adequate supply should be obtained in a normal well balanced diet.

Functions

1. It is essential for the hardening of bones and teeth, therefore an adequate supply in young people is important.
2. It plays an important part in the coagulation of blood.
3. It is associated with the process of contraction of muscles.

PHOSPHORUS

Sources of phosphorus include cheese, oatmeal, liver and kidney. If there is sufficient calcium in the diet there will be no deficiency of phosphorus.

Functions

It combines with calcium in the hardening of bones and teeth and helps to maintain the constant composition of the body fluids.

SODIUM

Sodium is supplied to the body in fish, meat, eggs, milk, and in the form of sodium chloride as table salt. Normal intake of sodium chloride per day varies from 5 to 20g and the normal daily requirement is between 2 and 5g per day. Excess is excreted in the urine.

Functions

1. It plays an important part in all cell activity.
2. A constant concentration is present in all cells and tissue fluids.

POTASSIUM

This substance is to be found widely distributed in all food. It plays an important part in the maintenance of the osmotic equilibrium. Normal intake of potassium chloride varies from 5 to 7g per day.

IRON

Iron, as a soluble compound, is found in liver, kidney, beef, egg yolk, whole meal bread and green vegetables. In normal adults about 1 mg iron is lost from the body daily. The usual daily diet contains about 20 mg but the amount absorbed is limited to the amount lost.

Functions

1. Iron is essential for the formation of haemoglobin in the red blood cells.
2. It is necessary for tissue oxidation.

IODINE

Iodine is found in salt water fish and in vegetables which have grown in soil containing iodine. In some parts of the world where iodine is deficient in the soil it is added in very small quantities to table salt. The daily requirement of iodine depends upon the individual's metabolic rate. Some people have a higher normal metabolic rate than others and their iodine requirements are greater. The minimum daily requirement is 20 μg.

Function

It is essential for the formation of *thyroxine* and *triiodothyronine*, the hormones secreted by the thyroid gland.

Roughage

Roughage is the undigestable part of the diet, for example, cellulose of fruit and vegetables and connective tissue in protein foods.

Functions

1. It gives bulk to the diet and so helps to satisfy the appetite.
2. It stimulates peristalsis, that is, the muscular activity of the alimentary tract.
3. It stimulates bowel movement.

Water

Water is a liquid compound of hydrogen and oxygen, made up by the chemical combination of two parts of hydrogen and one part of oxygen (H_2O). Water makes up about 60 per cent of the body weight in men and about 50 per cent in women.

Functions

1. It provides the moist environment which is required by all living cells in the body.
2. It participates in all the chemical reactions which occur inside and outside the body cells.
3. It dilutes and moistens food.

4. It assists in the regulation of body temperature by the evaporation of sweat from the skin.

5. As a major component of blood and tissue fluid it transports water soluble substances round the body.

6. It dilutes waste products and poisonous substances in the body.

7. It contributes to the formation of urine and faeces.

For tables summarising vitamins see pages 226–228.

Summary—Fat Soluble Vitamins

Letter	Chemical name	Source	Stability	Functions	Deficiency diseases	Daily requirement
A	Carotene (pro-vitamin)	Milk, butter, cheese, egg-yolk, fish, liver, oils, green and yellow vegetables	Some loss at high temperatures and long exposure to light and air	Maintains healthy epithelial tissues and cornea. Formation of visual purple	Keratinisation Xerophthalmia Stunted growth Night blindness	5000 I.U.
D	Choli-calciferol and calciferol	Fish, liver, oils, milk, cheese and egg-yolk, irradiated 7-dehydrocholesterol in human skin	Very stable	Facilitates the absorption and utilisation of calcium and phosphorus = healthy bones and teeth	Rickets Osteomalacia	400–800 I.U.
E	Tocopherol	Egg-yolk, milk, butter, green vegetables, nuts	Destroyed by rancid fat and iron salts	Maintains healthy muscular system	Very rare muscular dystrophies	Unknown
K	Menaphthone	Leafy vegetables, fish, liver, fruit	Destroyed by light, strong acids and alkalis	Formation of prothrombin in the liver	Slow blood clotting Haemorrhages in the newborn	

Bile is necessary for the absorption of these vitamins.
Mineral oils interfere with absorption.

Summary—Water Soluble Vitamins

Letter	Chemical name	Source	Stability	Functions	Deficiency diseases	Daily requirement
C	Ascorbic acid	Citrus fruits, currants, berries, green vegetables, potatoes, liver and glandular tissue in animals	Destroyed by heat, aging, acids, alkalis, chopping, salting and drying	Formation of intercellular matrix. Maturation of red blood cells	Multiple haemorrhages. Slow wound healing. Anaemia. Gross deficiency causes scurvy	30–60 mg
B₁	Thiamin	Yeast, liver, germ of cereals, nuts, pulses, rice polishings, egg yolk, liver, legumes	Stable	Metabolism of carbohydrates and nutrition of nerve cells. Efficient water exchange in the body	General fatigue and loss of muscle tone. Ultimately leads to beri beri	1 mg
B	Niacin	Yeast, offal, fish, pulses, whole meal cereals. Synthesised in the body from tryptophan	Fairly stable	Necessary for tissue oxidation	Prolonged deficiency causes pellagra, i.e. dermatitis diarrhoea dementia	10 mg approx.
B₂	Riboflavin	Liver, yeast, milk, eggs, green vegetables, kidney	Destroyed by light and alkalis	Necessary for tissue oxidation and growth	Angular stomatitis Cheilosis Dermatitis Eye lesions	1–2 mg
B₁₂	Cyanocobalamin	Liver, milk, moulds, fermenting liquors, egg	Destroyed by heat	Maturation of R.B.C.s	Pernicious anaemia Degeneration of nerve fibres of the spinal cord	Unknown

Summary—Water Soluble Vitamins—continued

Letter	Chemical name	Source	Stability	Functions	Deficiency diseases	Daily requirement
B	Folic acid	Dark green vegetables, liver, kidney, eggs synthesised in colon	Destroyed by heat and moisture	Formation of R.B.C.s	Anaemia	Unknown
B₆	Pyridoxine	Meat, liver, vegetables, bran of cereals, egg-yolk, beans, soya beans	Stable	Protein metabolism. Formation of R.B.C.s and W.B.C.s	Very rare	Unknown
B	Pantothenic acid	Liver, yeast, egg-yolk, fresh vegetables	Destroyed by excessive heat and freezing	Probably required for formation of R.B.C.s	'Burning feet' syndrome	Unknown
B	Biotin	Yeasts, liver, kidney, pulses, nuts	Stable	Carbohydrate and fat metabolism. Growth of bacteria	Dermatitis conjunctivitis, mental disturbances	Unknown
B	Choline	Egg-yolk, liver pulses	Destroyed by alkalis	Transport of fatty acids	Fatty degeneration of the liver	Unknown

Para amino benzoic acid and Inositol are two vitamins of the B group about which little is known. Hesperidin or vitamin P is related to vitamin C. It is found in citrus fruits, berries, currants, etc., and it is thought to play some part in capillary resistance.

11. The Digestive System

The digestive system is the system which deals with the food which we eat. As food passes through the digestive tract it is broken down by *physical* and *chemical means,* until it is in a form suitable for absorption into the blood stream and utilisation within the body. There are certain constituents of the diet which cannot be digested and absorbed. therefore they are excreted in the form of faeces. The digestive system performs four major activities.

1. *Ingestion.* This means taking food into the alimentary tract.

2. *Digestion.* Physical digestion consists of mechanical breakdown of food by *mastication* (chewing), and *muscular action* within the digestive tract. Chemical digestion is achieved by chemical substances or *enzymes** present in secretions produced by glands of the digestive system, these secretions include saliva from the salivary glands, gastric juice from the stomach, intestinal juice from the small intestine, bile from the liver and pancreatic juice from the pancreas.

3. *Absorption.* This is the process by which digested food substances pass through the walls of some organs or the digestive tract into the blood and lymph capillaries.

4. *Elimination.* Food substances which have been eaten but cannot be absorbed are excreted from the bowel as faeces.

The Organs of the Digestive System
The alimentary tract
This is a long tube through which food passes. It commences at the mouth and terminates at the anus and the various parts are given separate names, although structurally they are remarkably similar.

*An enzyme is a chemical substance which causes or speeds up a chemical change in other substances, without itself being changed.

The parts of the tract are:
    mouth
    pharynx
    oesophagus
    stomach
    small intestine
    large intestine
    rectum and anus.

Figure 187

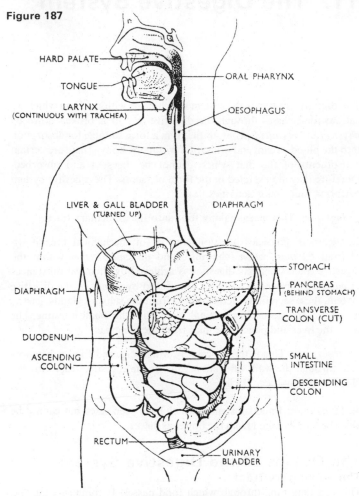

HARD PALATE

TONGUE

LARYNX
(CONTINUOUS WITH TRACHEA)

ORAL PHARYNX

OESOPHAGUS

LIVER & GALL BLADDER
(TURNED UP)

DIAPHRAGM

DIAPHRAGM

STOMACH

PANCREAS
(BEHIND STOMACH)

TRANSVERSE
COLON (CUT)

DUODENUM

ASCENDING
COLON

SMALL
INTESTINE

DESCENDING
COLON

RECTUM

URINARY
BLADDER

General view of the organs of the digestive system.

Accessory organs

Various secretions are poured into the alimentary tract, some by
glands in the lining membrane of the organs, for example, gastric juice

by the lining of the stomach; and some by glands situated outside the tract. The latter are the accessary organs of digestion and their secretions pass through ducts to enter the tract. They consist of:

three pairs of salivary glands

pancreas

liver and the biliary tract.

MOUTH OR ORAL CAVITY

The mouth is a cavity bounded by muscles and bones.

Anteriorly. By the muscles of the lips.

Posteriorly. It is in communication with the pharynx.

Laterally. By the muscles of the cheeks.

Superiorly. By the bony hard and muscular soft palate.

Inferiorly. By the muscular tongue and the muscles and soft tissues of the floor of the mouth.

The oral cavity is lined throughout with *mucous membrane* which consists of *stratified squamous epithelium* containing small mucus secreting glands.

The part of the mouth between the cheeks and lips, and the gums and teeth, is called the *vestibule* and the remainder of the cavity the *mouth proper*. The mucous membrane lining of the cheeks and the lips is reflected on to the gums or alveolar ridges.

Figure 188

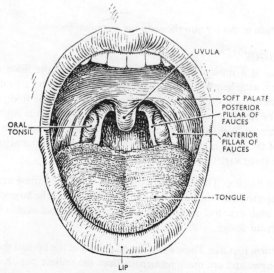

Structures seen in the widely open mouth.

The palate is divided into the anterior part which is called the *hard palate* and the posterior part called the *soft palate*. The bones forming the hard palate are the superior maxillae and the palatine bones. The soft palate is muscular, curves downwards from the posterior end of the hard palate and blends with the walls of the pharynx at the sides.

The *uvula* is a curved fold of muscle covered with mucous membrane which hangs down from the middle of the free border of the soft palate. Originating from the upper end of the uvula there are four folds of mucous membrane, two passing downwards at each side to form the *anterior and posterior pillars of the fauces.* Between each pair of pillars there is a collection of lymphoid tissue called the *oral tonsil.*

Figure 189

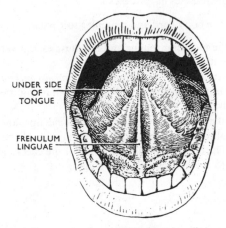

Under-surface of the tongue showing the frenulum linguae.

THE TONGUE

The tongue is a muscular structure which occupies the floor of the mouth. It is attached by its base to the *hyoid bone* and by a fold of its mucous membrane covering, called the *frenulum,* to the floor of the mouth. The superior surface consists of stratified squamous eqithelium with numerous *papillae* (little projections) which contain the nerve endings of the sense of taste, these are sometimes called the *taste buds.* There are three varieties of papillae.

1. *Vallate Papillae.* There are usually about 8 to 12 of these arranged in a V shape towards the base of the tongue. These are the largest of the papillae and are the most easily distinguished because of their characteristic design.

2. *Fungiform Papillae.* These are situated mainly at the tip and the edges of the tongue and are more numerous than the vallate papillae.

3. *Filiform Papillae.* These are the smallest of the three types and are

found to be most numerous on the edges and the anterior two-thirds of the tongue.

Figure 190

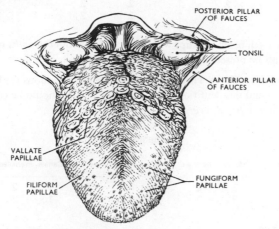

POSTERIOR PILLAR OF FAUCES

TONSIL

ANTERIOR PILLAR OF FAUCES

VALLATE PAPILLAE

FILIFORM PAPILLAE

FUNGIFORM PAPILLAE

Diagram showing the papillae of the tongue.

Functions of the tongue

The tongue plays an important part in mastication (chewing), deglutition (swallowing) and in speech. It is the organ of taste and the nerve endings of the sense of taste are to be found in the papillae.

THE TEETH

The teeth are embedded in the alveoli or sockets of the alveolar ridges of the mandible and the maxillae. Each individual has two sets of teeth, the *temporary or deciduous teeth* and the *permanent teeth*. At birth the teeth of both dentitions are present in immature form in the mandible and maxillae.

Figure 191

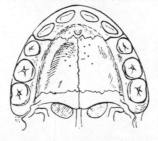

Upper jaw of a child showing the deciduous teeth.

The Temporary Teeth. These are 20 in number, 10 in the upper jaw and 10 in the lower jaw.

Deciduous Teeth

Jaw	Molars	Canine	Incisors	Incisors	Canine	Molars
Upper	2	1	2	2	1	2
Lower	2	1	2	2	1	2

These teeth begin to erupt when the child is about *6 months* old, and should all be present by the end of *24 months*.

The Permanent Teeth. These begin to replace the deciduous teeth in the *sixth year* and this dentition, consisting of 32 teeth, should be complete by the *twenty-fourth year*.

Permanent teeth

Jaw	Molars	Premolars	Canine	Incisors	Incisors	Canine	Premolars	Molars
Upper	3	2	1	2	2	1	2	3
Lower	3	2	1	2	2	1	2	3

Figure 192

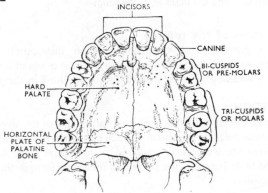

Permanent teeth in the upper jaw, as seen from below.

Figure 193

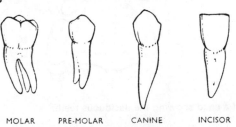

MOLAR   PRE-MOLAR   CANINE   INCISOR

Diagram showing roots of the permanent teeth.

The *incisor* and *canine* teeth are the cutting teeth and are used for biting off pieces of food; whereas the *premolar* and *molar* teeth, which have a broad, flat surface, are used for grinding or chewing the food. The incisors and canine teeth have only one root, the pre-molars two and the molars three roots.

Structure of a tooth

Although the shape of the different teeth vary the structure is the same and consists of:

the crown is the part which protrudes from the gum

the root is the part embedded in the bone

the neck is the slightly constricted part where the crown merges with the root.

Figure 194

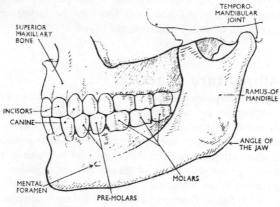

The teeth—side view.

Figure 195

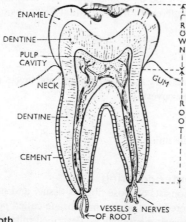

Section of a tooth.

In the centre of the tooth there is the *pulp cavity* containing blood vessels, lymph vessels and nerves and surrounding this there is a hard ivory-like substance called *dentine*. Outside the dentine of the crown of the tooth there is a thin layer of very hard substance called *enamel*. The root of the tooth, on the other hand, is covered with a substance resembling bone, called *cement*, which fixes the tooth in its socket. There is a small foramen at the apex of the root of the tooth which allows for the passage of blood vessels and nerves to and from the tooth.

THE PHARYNX

As has already been described the pharynx is divided for descriptive purposes into three parts, the naso-pharynx, the oro-pharynx and the laryngo-pharynx. Of these the *oro-pharynx* and the *laryngo-pharynx* are associated with the alimentary tract. Food passes from the oral cavity to the oro-pharynx and from there to the laryngo-pharynx which is continuous with the *oesophagus* below.

The Alimentary Tract

For descriptive purposes the alimentary tract is divided into a number of parts or organs, i.e. oesophagus, stomach, small intestine, large intestine, rectum and anus. Some of the functions of the tract are common to all its parts, e.g., the onward movement of the ingested food, whereas other activities are carried out exclusively by one or possibly two organs, e.g., the absorption of the products of digestion.

The structure of the alimentary tract may be described in relation to these two types of function. There is a common structural plan throughout the length of the tract which may be modified in an organ because of its functional specialisation.

The general plan will be described first, then the variations in this plan which are associated with the individual functions of the different organs.

GENERAL PLAN

The wall of the alimentary tract consists of four layers of tissue:
    adventitia or outer covering
    muscle layer
    submucous layer
    mucous membrane lining.

Adventitia or outer covering

In the thorax this consists of loose fibrous tissue, and in the abdomen the organs are covered by a serous membrane called *peritoneum*. The arrangement of the peritoneum will be described in more detail later.

Muscle layer

With some exceptions this consists of two layers of smooth muscle. The muscle fibres of the outer layer are arranged longitudinally, and those of the inner layer encircle the wall of the tract. Between these two muscle layers there is a plexus of nerves, called the *myenteric or Auerbach's plexus,* which contains sympathetic and parasympathetic nerves.

The contraction of these muscle layers occurs in waves which push the contents of the tract onwards. This type of contraction of smooth muscle is called *peristalsis.*

Submucous layer

This layer consists of loose connective tissue with some elastic fibres. Within this layer there are lymph vessels and plexuses of blood vessels and nerves. The blood vessels consist of arterioles, venules and capillaries. The nerve plexus is called *Meissner's plexus* and it contains sympathetic and parasympathetic nerves which supply the mucous membrane lining.

Mucous membrane

This layer has three main functions: protective, secretory and absorptive. In parts of the tract which are subject to mechanical injury this layer consists of stratified squamous epithelium with mucus secreting glands just below the surface. Where secretion of digestive juices and absorption occur the mucous membrane consists of columnar epithelial cells through which absorption occurs, goblet cells which secrete mucus and glands made up of special cells which secrete digestive juices containing enzymes. Under the epithelial lining there are varying amounts of lymphoid tissue.

NERVE SUPPLY

The alimentary tract is supplied by nerves from both parts of the autonomic nervous system, i.e. parasympathetic and sympathetic, and in the main their actions are antagonistic. In the normal healthy state one influence may outweigh the other according to the needs of the body as a whole at any point in time.

The parasympathetic supply, which is provided by two cranial nerves, the vagus nerves, stimulates muscle contraction and the secretion of digestive juices.

The sympathetic supply is provided by numerous nerves which emerge from the spinal cord in the thoracic and lumbar regions. These nerves form plexuses in the thorax, abdomen and pelvis, and from them nerves pass to the organs of the alimentary tract. Their action is to reduce muscle contraction and glandular secretion.

THE OESOPHAGUS

The oesophagus or gullet is about 10 inches long and is the narrowest part of the alimentary tract. It lies in the median plane in front of the vertebral column and behind the trachea and the heart. It is continuous with the pharynx above and passes through the mediastinum and the central tendon of the diaphragm to the stomach. Immediately the oesophagus passes through the diaphragm it curves upwards before joining the stomach. This change of direction of the oesophagus is believed to be one of the factors which prevents the regurgitation of gastric contents into the oesophagus.

STRUCTURE

There are four layers of tissue which form the oesophagus.

1. The outer covering consists of *elastic fibrous tissue*

2. The *muscle tissue* consists of a longitudinal and a circular layer of fibres. In the upper third the muscle fibres are striated but their action is involuntary. Gradually the striated fibres are replaced by smooth muscle fibres. At the lower end of the oesophagus the circular fibres are thickened to become the *cardiac sphincter.**

3. The *submucous layer* which consists of *areolar tissue*, blood vessels, lymph vessels and nerves of the autonomic nervous system.

4. The inner lining consists of *stratified squamous epithelium* with mucus secreting glands which have tiny ducts leading to the surface.

BLOOD SUPPLY

Arterial

Branches from the *thoracic aorta* and the *gastric arteries*.

Venous

Azygous vein
Left gastric vein.

NERVE SUPPLY

Sympathetic and parasympathetic nerves.

THE STOMACH

The stomach is a J-shaped dilated portion of the alimentary tract situated in the epigastric, umbilica and left hypochondriac regions of the abdominal cavity.

Sphincter muscle is the name given to a thickened area of circular involuntary muscle fibres which contract and close an orifice.

Figure 196

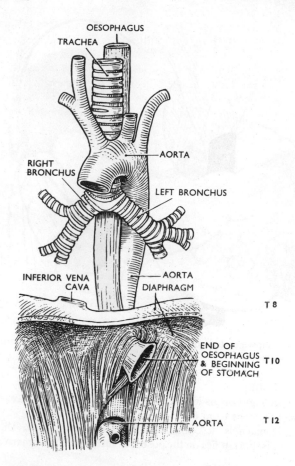

Diagram showing the organs in association with the oesophagus.

Organs in association with the stomach

Anteriorly. The left lobe of the liver and the anterior abdominal wall.

Posteriorly. The abdominal aorta, the inferior vena cava, the vertebral column, the diaphragm, the pancreas and the spleen.

Superiorly. The diaphragm, the oesophagus and the left lobe of the liver.

Inferiorly. The transverse colon and the small intestine.

To the Left. The left kidney and the spleen.

To the Right. The liver and the duodenum.

Figure 197

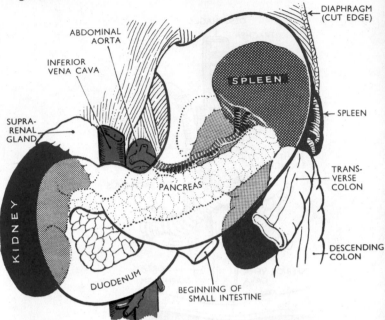

Diagram showing the position of the stomach and the organs in association with it.

STRUCTURE

The oesophagus opens into the stomach at the *cardiac orifice*, and the duodenum opens from the stomach at the *pyloric orifice*.

The stomach is described as having two curvatures. The *lesser curvature* is short, it lies on the posterior surface of the stomach and is a

Figure 198

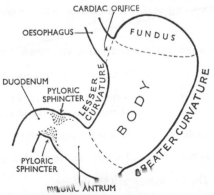

Scheme of the stomach showing the fundus, body and pyloric antrum.

continuation downwards of the posterior part of the oesophagus. Just before the pyloric sphincter it curves upwards to complete the J shape. The *greater curvature* is on the anterior surface of the stomach. Where the oesophagus enters the stomach the anterior part angles acutely upwards, curves downwards then slightly upwards towards the pyloric sphincter.

The part of the stomach above and to the left of the cardiac orifice is called the *fundus*, the main part is the *body* and the lower part which curves to the right is the *pyloric antrum*. At the distal end of the pyloric antrum there is a sphincter, the *pyloric sphincter*, which guards the opening between the *pyloric antrum* and the duodenum.

There are four layers of tissue which form the walls of the stomach.

Figure 199

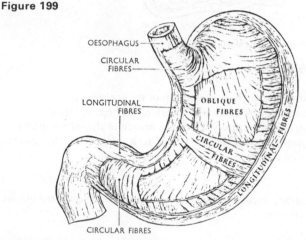

Diagram illustrating the arrangement of muscle fibres of the stomach wall.

1. The outer *serous covering* which consists of peritoneum.

2. The muscle layer lying immediately deep to the serous coat has three layers of smooth muscle fibres. The outer layer has *longitudinal fibres*, the middle layer has *circular fibres* and the inner layer, *oblique fibres*. This arrangement allows for the churning motion characteristic of gastric activity.

3. The *submucous layer* consists of areolar tissue, blood and lymph vessels and nerves of the autonomic nervous system.

4. *The mucous membrane lining*. When the stomach is empty the mucous membrane lining is thrown into longitudinal folds or *rugae*, and when it is full the rugae are 'ironed out' and the surface has a smooth, velvety appearance. The mucous membrane consists of columnar epithelium with goblet cells. Within the mucous membrane there are numerous gastric glands which secrete gastric juice.

Gastric juice contains enzymes which assist in the digestion of protein foods.

Fig. 200

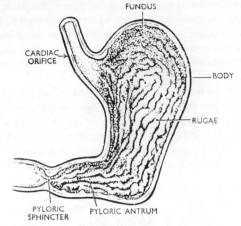

Interior of the stomach showing rugae.

BLOOD SUPPLY

Arterial

Left gastric artery which is a branch of the coeliac artery.

Right gastric artery which is a branch of the hepatic artery.

Left gastro-epiploic artery which is a branch of the splenic artery.

Right gastro-epiploic artery which is a branch of the gastro-duodenal artery.

Venous

A similar arrangement of veins conveys venous blood to the tributaries of the portal vein.

NERVE SUPPLY

Sympathetic and parasympathetic.

THE SMALL INTESTINE

The small intestine is continuous with the stomach at the *pyloric sphincter* and leads into the large intestine at the *ileocolic valve*. It is about 21 feet in length and lies in the abdominal cavity surrounded by the large intestine (Fig. 203). In the small intestine the chemical digestion of food is completed and most of the absorption of nutrient materials takes place.

For descriptive purposes it is divided into three parts which are continuous with each other.

Duodenum

The duodenum is about 10 inches in length and curves in a C around the head of the pancreas. At the mid-point of the duodenum there is the common opening of the pancreatic duct and the bile duct which is guarded by the *sphincter of Oddi*.

Figure 201

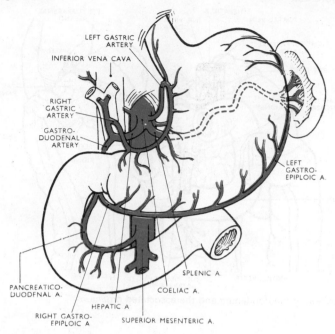

Arterial blood supply to the stomach.

Jejunum
The jejunum is the middle part of the small intestine and is about 8 feet in length.

Ileum
The ileum is about 12 feet long and terminates at the *ileocolic valve* which controls the flow of material from the ileum to the large intestine and vice versa.

STRUCTURE
There are four layers of tissue forming the walls of the small intestine.

1. The outer covering of *peritoneum* called the *mesentery*.

2. *The muscle layer*. There are two layers of smooth muscle fibres under the serous membrane; a superficial layer of longitudinal and a deeper layer of circular fibres.

3. *The submucous layer* consists of areolar tissue, blood vessels, lymph vessels and nerves of the autonomic nervous system.

Figure 202

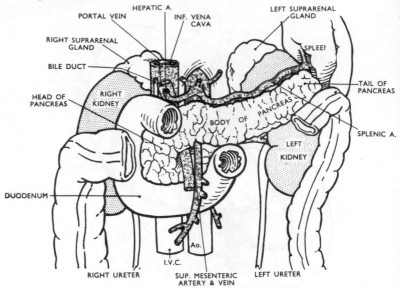

Position of the duodenum and the associated organs.

Figure 203

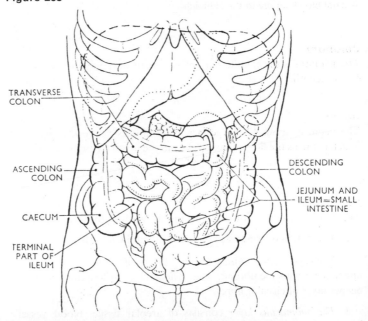

Diagram showing the position of the small intestine.

4. *The mucous membrane* lining. The surface area of the small intestine is greatly increased by two peculiarities in the arrangement of the mucous membrane.

a. *The circular folds,* unlike the rugae of the stomach, are not smoothed out when the small intestine is distended.

b. *The villi,* which are tiny finger-like projections into the lumen of the organ. Their walls consist of columner epithelial cells which enclose a network of blood and lymph capillaries. The lymph capillaries are called *lacteals.* Absorption of nutrient materials takes place across the wall of the villus into the blood and lymph capillaries.

Figure 204

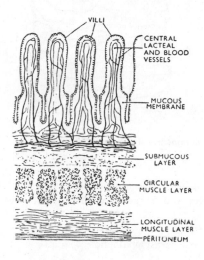

Illustration showing magnified view of the walls of the small intestine.

Intestinal Glands. These are simple tubular glands which lie between the villi. They secrete *intestinal juice* containing the enzymes which complete the chemical digestion of carbohydrates, proteins and fats.

Lymph Nodes. There are numerous lymph nodes, in the mucous membrane situated at irregular intervals throughout the length of the small intestine. The smaller ones are known as *solitary lymphatic nodules,* and about 20 or 30 larger nodules situated towards the distal end of the ileum are called *aggregated lymphatic nodules* (Peyer's patches).

BLOOD SUPPLY
Arterial
The *superior mesenteric artery* which is a branch of the abdominal aorta.

Figure 205

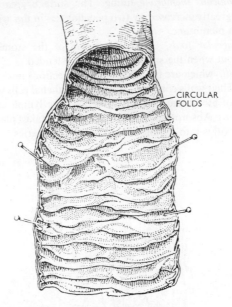

CIRCULAR
FOLDS

Interior of small intestine showing circular folds.

Venous

The *superior mesenteric vein* which units with the inferior mesenteric and splenic veins and the veins from the stomach to form the *portal vein*.

NERVE SUPPLY

Sympathetic and parasympathetic.

Figure 206

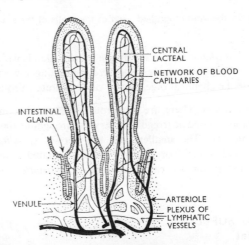

CENTRAL
LACTEAL

NETWORK OF BLOOD
CAPILLARIES

INTESTINAL
GLAND

VENULE

ARTERIOLE

PLEXUS OF
LYMPHATIC
VESSELS

Highly magnified view of two villi and intestinal glands.

Figure 207

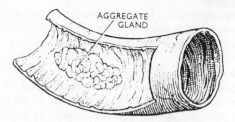

AGGREGATE
GLAND

Interior of a section of the small intestine showing an aggregate gland.

THE LARGE INTESTINE OR COLON

The large intestine is about *five feet* long, beginning at the *caecum* in the right iliac fossa and terminating at the *rectum* and *anal canal* deep in the pelvis. Its lumen is larger than that of the small intestine. It forms an arch round the coiled up small intestine (Fig. 209).

For descriptive purposes it is divided into a number of parts.

Caecum

This is the first part of the colon. It is a dilated portion which has a blind end inferiorly and is continuous with the *ascending colon* superiorly. Just below the junction of the two the ileocolic valve opens from the ileum on the medial aspect of the caecum. The *vermiform appendix* is a fine tube, closed at one end, which leads from the caecum. It is usually about five inches long and has the same structure as the walls of the colon but contains more lymphoid tissue.

Figure 208

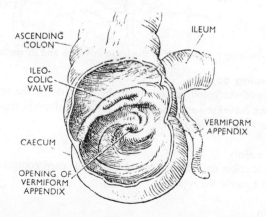

ASCENDING
COLON

ILEO-
COLIC
VALVE

CAECUM

OPENING OF
VERMIFORM
APPENDIX

ILEUM

VERMIFORM
APPENDIX

Interior of the caecum showing the ileo-colic valve and the opening of the vermiform appendix.

Ascending colon

The ascending colon passes upwards from the caecum to the level of the liver where it bends acutely to the left at the *right colic flexure* (hepatic flexure) to become the *transverse colon*.

Transverse colon

This is a loop of colon which extends transversely across the abdominal cavity in front of the duodenum and the stomach to the spleen where it forms the *left colic flexure* (splenic flexure) by bending acutely downwards to become the *descending colon*.

Figure 209

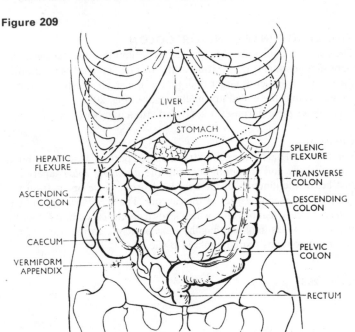

Diagram showing the position of the large intestine and its parts.

Descending colon

This part of the colon passes down the left side of the abdominal cavity then curves towards the midline. After it enters the true pelvis it is known as the *pelvic colon*.

Pelvic colon

The pelvic colon describes a loop in the pelvis then continues downwards to become the *rectum*.

Rectum

This is a slightly dilated part of the colon which is about *five inches* long. It leads from the pelvic colon and terminates in the *anal canal*.

Anal canal

This is a short canal about *one inch* in length which leads from the rectum to the exterior. There are two sphincter muscles which control the anus; the internal sphincter which consists of smooth muscle fibres is under the control of the *autonomic nervous system* and the external sphincter, formed by striated muscle, is under *voluntary nerve* control.

Figure 210

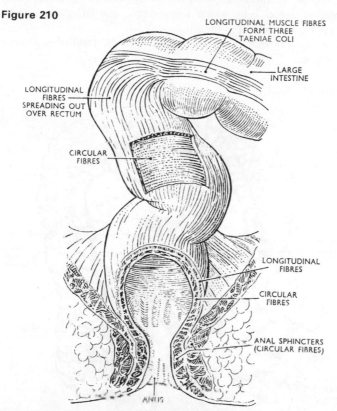

Diagram showing the arrangement of muscle fibres of the large intestine, rectum and anus.

STRUCTURE

1. The outer covering of *peritoneum*.

2. *The muscle layer*. There are two layers of smooth muscle fibres. *The longitudinal fibres* do not form a smooth continuous layer of tissue but are collected into three bands called *taeniae coli*, which are situated at regular intervals round the colon. As these bands of muscle tissue are slightly shorter than the total length of the colon they give a sacculated or puckered appearance to the organ (Fig. 210). In the

rectum the longitudinal fibres spread out to completely surround the organ. *The circular muscle fibres* form a thin layer which completely surrounds the colon. The anal canal sphincters are formed by thickening of these circular fibres.

3. *The submucous layer,* which consists of areolar tissue, blood vessels, lymph vessels, lymphoid tissue and nerves of the autonomic nervous system.

4. *The mucous membrane* lining consists of columnar epithelium with numerous goblet cells which form simple tubular glands. These glands are not present in the anus.

The lining membrane of the *anus* consists of stratified squamous epithelium.

Figure 211

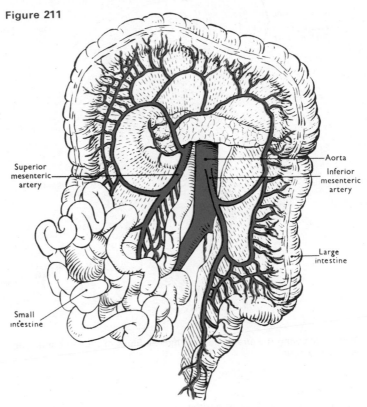

Arterial blood supply to the small and large intestines.

BLOOD SUPPLY

Arterial

The superior mesenteric artery supplies the caecum, ascending and transverse colon.

The inferior mesenteric artery supplies the remainder of the colon and the rectum.

Venous

The superior and inferior mesenteric veins which join with the splenic and gastric veins to form the portal vein.

NERVE SUPPLY

Sympathetic and parasympathetic with the exception of the external anal sphincter which is under voluntary control.

The Peritoneum

The peritoneum is the largest serous membrane of the body and consists of a closed sac within the abdominal cavity. It is described as having two layers.

Figure 212

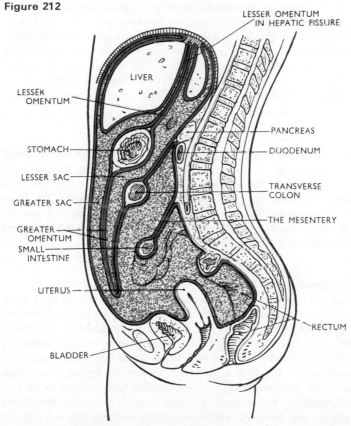

LESSER OMENTUM IN HEPATIC FISSURE

LIVER

LESSER OMENTUM

STOMACH

LESSER SAC

GREATER SAC

GREATER OMENTUM

SMALL INTESTINE

UTERUS

BLADDER

PANCREAS

DUODENUM

TRANSVERSE COLON

THE MESENTERY

RECTUM

Diagram showing the position of the peritoneum.

1. *The parietal layer* which lines the abdominal wall.

2. *The visceral layer* which covers the organs or viscera within the abdominal and pelvic cavities.

The arrangement of the peritoneum is complicated, it is as though the organs had been invaginated into it from below, behind and above. This means that the pelvic organs are covered only superiorly, the intestine is surrounded by it and is attached by a double layer to the posterior abdominal wall, and the liver is almost completely covered with peritoneum which attaches it to the under surface of the diaphragm. Most of the abdominal organs are invaginated into the peritoneum from behind, thus the parietal layer lines the anterior abdominal wall. The two layers of peritoneum are actually in contact and friction between them is prevented by the presence of serous fluid which is secreted by the cells, thus the *peritoneal cavity* is only a *potential cavity*.

In the male the peritoneal cavity is completely closed but in the female the uterine tubes open into it.

Some structures such as the ileum, jejunum, transverse and pelvic colons have long attachments to the posterior wall of the abdomen. This gives them greater freedom of movement to glide over each other. On the other hand some organs have a very short attachment or are covered only on their anterior surfaces, for example, the duodenum and ascending colon have a very short attachment, and the kidneys, pancreas and spleen have peritoneum only on their anterior surfaces.

The mesentery

This is the name given to the double layer of peritoneum which encloses the ileum and jejunum attaching them to the posterior abdominal wall. The attachment is quite short in comparison with the length of the small intestine, therefore it is fan shaped. The large blood vessels and nerves lie on the posterior abdominal wall and the branches from them to the small intestine pass between the two layers of the mesentery.

The greater omentum

This is a long loose fold of peritoneum originating from the greater curvature of the stomach, and hangs down in front of the abdominal organs like an apron. It is then reflected to the posterior abdominal wall. The greater omentum has a considerable store of fat within its folds and it is richly supplied with blood vessels, lymphatic vessels and lymph nodes. It acts as a protection against the spread of infection within the peritoneal cavity, by surrounding and isolating an area of inflammation, for example, in chronic appendicitis the vermiform appendix may be completely surrounded by the omentum and isolated from the rest of the abdominal organs.

The lesser omentum

This is the name given to a fold of peritoneum which extends from the inferior surface of the liver to the lesser curvature of the stomach. Behind the stomach there is a pocket of peritoneum called the *lesser sac,* while the remainder of the peritoneal cavity is called the *greater sac.*

The transverse mesocolon

This is a double fold of peritoneum which suspends the transverse colon from the posterior abdominal wall. The arrangement of blood vessels and nerves is the same as those to the small intestine.

The pelvic mesocolon

This is the double fold of peritoneum which suspends the pelvic colon.

Figure 213

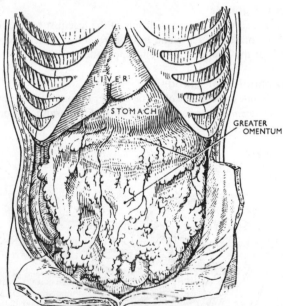

Diagram illustrating the greater omentum.

Functions of the peritoneum

1. It gives support to the abdominal organs.

2. Due to the presence of serous fluid secreted by its cells, the organs can glide smoothly over each other without causing friction.

3. It acts as a protection against infection due to the presence of lymph nodes and the ability of the greater omentum to isolate an area of inflammation.

4. It forms a protective covering for the abdominal organs.

5. It acts as a store for fat.

The Accessory Glands of Digestion

These are glands situated outside the alimentary tract. Their function is to secrete digestive juices which are conveyed to the tract by ducts. They are:

the salivary glands
the pancreas
the liver.

THE SALIVARY GLANDS

There are three pairs of *compound racemose glands* which pour their secretions into the mouth. They are:

2 parotid
2 submandibular
2 sublingual.

Parotid glands

These are situated one on each side of the face just below the external auditory meatus. Each gland has a *parotid duct* which opens into the mouth at the level of the second upper molar tooth.

Submandibular glands

These lie one on each side of the face under the angle of the jaw. The two *submandibular ducts* open on to the floor of the mouth, one on each side of the frenulum linguae.

Figure 214

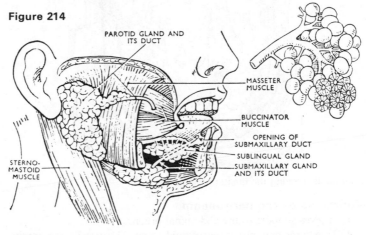

Diagram showing the positions of the salivary glands. Inset shows the minute structure of part of a salivary gland.

Sublingual glands

These glands lie under the mucous membrane of the floor of the mouth in front of the submandibular glands. They have numerous small ducts which pierce the mucous membrane of the floor of the mouth.

STRUCTURE OF THE SALIVARY GLANDS

These glands are all surrounded by a *fibrous capsule*. They consist of a number of *lobules* which are made up of small alveoli lined with *secretory cells*. The secretions are poured into small ducts which join up to form larger ducts leading into the mouth.

NERVE SUPPLY

Sympathetic and parasympathetic.

Saliva. This is the combined secretions from the salivary glands and the small mucus secreting glands of the lining of the oral cavity.

THE PANCREAS

The pancreas is a pale yellowish grey gland which weighs about 60 grammes (2 ounces). It is about 18 cm (7 inches) long, and is situated in the *epigastric* and the *left hypochondriac regions* of the abdominal cavity. It consists of a broad *head* which lies in the curve of the duodenum and is closely attached to it; a *body* which lies behind the body of the stomach and a narrow *tail* which lies in front of the left kidney and which just reaches the spleen. The abdominal aorta and the inferior vena cava pass behind the gland.

The pancreas consists of a number of *lobules* which are made up of small alveoli lined with *secretory cells*. Each lobule is drained by a tiny duct which unites with other ducts and joins the main *pancreatic duct*

Figure 215

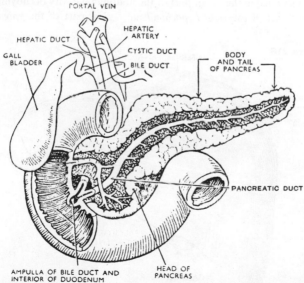

Diagram showing the position of the pancreas in relation to the duodenum and bile ducts.

which passes the whole length of the gland to open into the duodenum at its mid-point. Just before entering the duodenum the pancreatic duct joins the *bile duct* to form the *ampulla of the bile duct*. The duodenal opening is controlled by the *sphincter of Oddi*.

Islets of Langerhans

Distributed throughout the substances of the pancreas little collections of a different type of cell are to be found. These cells form what are known as the *islets of Langerhans*. The secretion produced by the islets of Langerhans is passed directly into the circulating blood and consists of the hormones *insulin* and *glucagon*.

Function

The function of the alveoli of the pancreas is to produce *pancreatic juice* which plays an important part in the chemical digestion of food.

BLOOD SUPPLY

The splenic and mesenteric arteries supply arterial blood to the pancreas and the venous drainage is by the veins of the same names.

NERVE SUPPLY

Sympathetic and parasympathetic.

THE LIVER

The liver is the largest gland in the body. It weighs about *three pounds* and is situated in the upper part of the abdominal cavity occupying the greater part of the *right hypochondriac region*, part of the *epigastric*

Figure 216

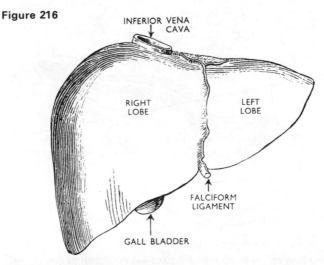

INFERIOR VENA CAVA

RIGHT LOBE

LEFT LOBE

FALCIFORM LIGAMENT

GALL BLADDER

The liver showing right and left lobes.

region and extending into the *left hypochondriac region*. Its upper and anterior surfaces are smooth and curved to fit the under surface of the diaphragm; its posterior surface is irregular in outline.

Organs in association with the liver

Superiorly and Anteriorly. Diaphragm and anterior abdominal wall.

Inferiorly. Stomach, bile ducts, duodenum, right colic flexure of the colon, right kidney and supra-renal gland.

Posteriorly. Oesophagus, inferior vena cava, aorta, gall bladder, vertebral column and the diaphragm.

Laterally. The lower ribs and the diaphragm.

Figure 217

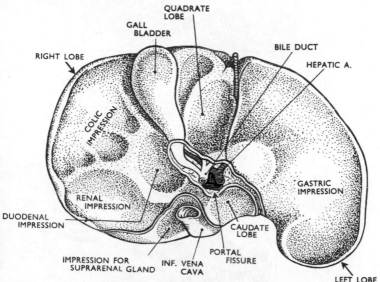

Diagram showing the posterior surface of the liver.

The liver is enclosed in a thin capsule and incompletely covered by a layer of peritoneum, folds of which form supporting ligaments to the diaphragm. It is described as having four lobes. The *right lobe* which is the largest, the *left lobe* which is smaller and wedge shaped, the *quadrate lobe* which is almost square in outline and the *caudate lobe* which is tail-like in appearance. These latter two can only be distinguished by viewing the liver from behind.

The portal fissure

This is the name given to the part on the posterior surface of the liver where various structures enter and leave the gland.

Figure 218

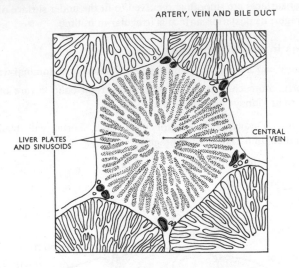

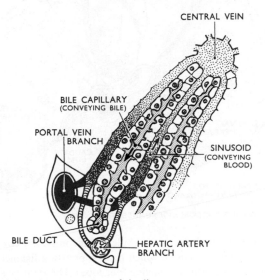

Diagram of the minute structure of the liver.

(a) A liver lobule.

(b) A schematic drawing to show the flow of blood and bile through a liver lobule.

The portal vein enters carrying blood from the stomach, spleen, pancreas and small and large intestine.

The hepatic artery enters carrying arterial blood. It is a branch from the coeliac artery which branches off from the abdominal aorta.

Nerve fibres, sympathetic and parasympathetic.

The right and left hepatic ducts leave carrying bile to the gall bladder.

Lymph vessels leave the liver at this point.

Figure 219

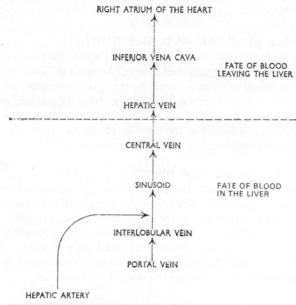

RIGHT ATRIUM OF THE HEART

INFERIOR VENA CAVA

FATE OF BLOOD
LEAVING THE LIVER

HEPATIC VEIN

CENTRAL VEIN

SINUSOID

FATE OF BLOOD
IN THE LIVER

INTERLOBULAR VEIN

PORTAL VEIN

HEPATIC ARTERY

Scheme of blood flow through the liver.

Venous drainage

The right and left hepatic veins emerge from the posterior surface of the liver and immediately enter the inferior vena cava just below the diaphragm.

STRUCTURE OF THE LIVER

The lobes of the liver are made up of tiny lobules which are just visible to the naked eye. These lobules are *hexagonal* in outline and are formed by cubical shaped cells arranged in columns which radiate from a *central vein*. Between the columns of cells there are *sinusoids* (blood vessels with incomplete walls) which contain a mixture of blood from the tiny branches of the portal vein and hepatic artery. This arrangement allows the oxygenated blood and the blood with a high concentration of nutritional materials to come into direct contact with the liver cells.

After the blood has been in contact with the liver cells it drains into the *central vein* or *intralobular vein*. Central veins from all the lobules join up, gradually becoming larger and larger until eventually they all unite to become the *hepatic veins* which drain blood from the liver as a whole and empty it into the inferior vena cava. Supporting the lobules of the liver there is lymphatic tissue.

One of the functions of the liver cells is to form *bile*. This is collected from the cells in very fine *bile capillaries* and drained away from them by larger bile ducts which eventually leave the liver as the *right and left hepatic ducts*.

THE GALL BLADDER AND BILE DUCTS

The right and left hepatic ducts join to form the *hepatic duct* just outside the portal fissure. The hepatic duct passes downwards for about an inch where it is joined at an acute angle by the *cystic duct* from the gall bladder. The cystic and hepatic ducts together form the *bile duct* which passes downwards posterior to the head of the pancreas to be joined by the main pancreatic duct at the ampulla of the bile duct. The two together open into the duodenum at the sphincter of Oddi.

STRUCTURE OF THE BILE DUCTS

The walls of the bile ducts consist of an outer *peritoneal covering, a smooth muscle layer* which exhibits peristaltic contraction and an inner lining of *mucous membrane*. In the cystic duct the lining membrane is arranged in irregularly situated circular folds which have the effect of a *spiral valve*. Bile passes through the cystic duct to the gall bladder. After it has been concentrated in the gall bladder it passes through the cystic duct again, then down the bile duct to the duodenum.

Figure 220

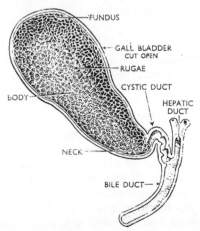

Interior of the gall bladder and bile ducts.

THE GALL BLADDER

The gall bladder is a pear-shaped sac attached to the under surface of
the liver by connective tissue. It is described as having a *fundus* or
expanded end, a *body* or main part and a *neck* which is continuous with
the cystic duct.

STRUCTURE OF THE GALL BLADDER

The surface of the gall bladder, not in contact with the liver, is covered
with *peritoneum*. Under the peritoneum there is a layer of smooth
muscle with longitudinal, circular and oblique fibres.

Figure 221

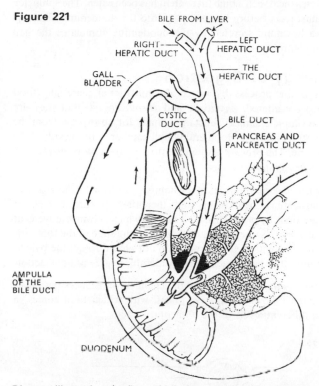

Diagram illustrating the flow of bile from the liver through the gall
bladder to the duodenum.

The inner lining is of *mucous membrane* which displays very small
rugae when the gall bladder is empty. The rugae disappear when the
organ is distended with bile.

BLOOD SUPPLY

The *cystic artery* which is a branch of the hepatic artery supplies blood
to the gall bladder. Blood is drained away by the *cystic vein* which joins
the portal vein.

NERVE SUPPLY

Nerve impulses are conveyed by sympathetic and parasympathetic nerve fibres.

FUNCTIONS

1. It acts as a reservoir for bile.
2. The lining membrane adds mucus to the bile.
3. By the absorption of water bile is concentrated in the gall bladder.
4. By the contraction of the muscular walls bile is expelled from the gall bladder and passed via the bile ducts into the duodenum. This occurs when a meal with a high fat content has been eaten. The sphincter of Oddi must relax before bile can pass into the duodenum. Cholecystokinin, a hormone secreted by the duodenum, stimulates the gall bladder to contract.

Physiology of Digestion

Digestion is the process by which the nutritional materials (food substances) are altered, *physically* and *chemically*, so that they are reduced to simple chemical substances ready for absorption from the alimentary tract into the blood and lymph capillaries in the villi of the small intestine. The process is divided into two main parts, *mechanical* and *chemical digestion.*

Mechanical digestion consists of the liquifying of food by the digestive juices, mastication, swallowing and thereafter onward movement through the tract by *peristalsis.* The rate at which onward movement takes place does, to some extent, depend upon the state of the food, for example, the contents of the stomach do not pass through the pyloric sphincter until they have reached a considerable degree of liquefaction.

Chemical digestion is effected by the chemical substances (*enzymes*) present in the following digestive juices, with which food comes in contact at different levels of the alimentary tract.

Figure 222

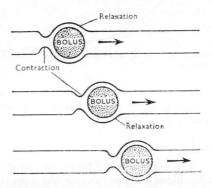

Illustration depicting peristalsis.

Saliva in the mouth.

Gastric juice in the stomach.

Pancreatic juice in the duodenum.

Bile in the duodenum.

Intestinal juice (succus entericus) in the small intestine.

Figure 223

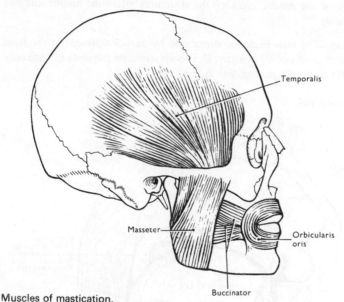

Muscles of mastication.

Digestion in the Mouth

When food is taken into the mouth it is masticated or chewed by the teeth and moved round the mouth by the tongue and by the muscles of mastication. It is moistened by saliva and formed into a soft mass or *bolus* ready for *deglutition* or swallowing.

SALIVA

Saliva is secreted into the mouth by the three pairs of salivary glands, and consists of:

    water

    mineral salts

    mucus added by the glands in the mouth

    enzyme; ptyalin or salivary amylase.

FUNCTIONS OF SALIVA

Digestion. The enzyme *ptyalin* acts on cooked starches (polysaccharides) changing them to *maltose*. This action is not completed in the mouth but continues in the stomach until the reaction of the bolus has been made strongly acid by hydrochloric acid in the gastric juice. The pH of saliva is about 6·7 and that of gastric juice between 1·5 and 1·8.

Lubrication of Food. Dry food entering the mouth must be moistened and lubricated before it can be made into a bolus ready for swallowing.

Cleansing and lubricating. An adequate flow of saliva is necessary to cleanse the mouth and keep the structures within the mouth soft and pliable.

Taste. The taste buds are stimulated by particles present in the food which are dissolved in water. Dry foods stimulate the sense of taste only after thorough mixing with saliva.

Figure 224

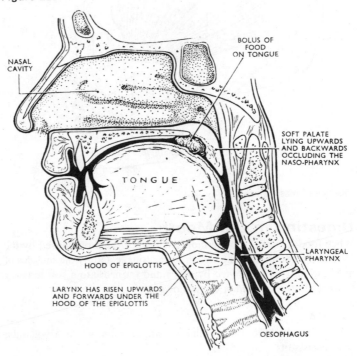

Side view of the buccal cavity showing the tongue, hard palate and soft palate, and movements which occur during swallowing.

SECRETION OF SALIVA

The flow of saliva is controlled by sympathetic and parasympathetic nerve supply. The autonomic control of salivation occurs in two ways.

Unconditioned reflex response to the presence of an object, such as food, in the mouth. This response may be elicited immediately after birth.

Conditioned reflex response is something which has been learned from previous experience. The sight, smell and even the thought of appetising food results in salivation sometimes called 'mouth watering'. This type of salivation occurs on the anticipation of food.

DEGLUTITION OR SWALLOWING
This occurs in three stages after mastication is complete and the bolus has been formed. After the process has been voluntarily initiated it is under autonomic nerve control.

1. The bolus is pushed backwards into the pharynx by the upward movement of the tongue.

2. The muscles of the pharynx propel the bolus down into the oesophagus. All other routes which the bolus could possibly take are closed. The soft palate rises up and occludes the naso-pharynx, the tongue and the pillars of the fauces close the way back into the mouth and the larynx is lifted upwards and forwards so that its opening is occluded by the overhanging epiglottis.

3. The presence of the bolus in the pharynx stimulates a wave of peristalsis which propels the bolus through the oesophagus to the stomach.

Digestion in the Stomach
The size of the stomach varies with the amount of food which it contains. When a sizable meal has been eaten the food accumulates in the stomach in layers, the last part remaining in the fundus for some time. Mixing with gastric juice takes place gradually and it may be some time before the food is sufficiently acidified to stop the action of ptyalin.

Peristaltic action in the stomach consists of a *churning* movement which is brought about by contraction of the three layers of muscle tissue. This churning movement causes further mechanical breakdown of the food, the mixing of the food with gastric juice and its onward movement into the duodenum.

GASTRIC JUICE
This is secreted by special secretory glands in the walls of the stomach and consists of:
    water
    mineral salts
    mucus
    hydrochloric acid
    enzymes; pepsinogen and rennin
    intrinsic factor.

FUNCTIONS OF GASTRIC JUICE

1. Because of its watery consistency it further liquefies the food swallowed.

2. The *hydrochloric acid* acidifies the stomach contents and terminates the action of ptyalin. *Pepsinogen* is converted to the active enzyme *pepsin* by hydrochloric acid.

The hydrochloric acid is believed to act as a barrier to the passage of certain micro-organisms which could be harmful to the body.

3. Enzyme action.

Pepsin begins the chemical digestion of proteins by converting them to peptones. It acts most effectively within the pH range of 1·6 to 3·2.

Rennin curdles milk by changing soluble caseinogen into insoluble *casein,* which in turn is coverted by pepsin into peptones. Rennin is present in the gastric juice of infants but is not present in adults.

4. Gastric juice contains the *intrinsic factor* which is necessary for the absorption of the *anti-anaemic factor* (cyanocobalamine, vitamin B_{12}). The anti-anaemic factor, present in food, is absorbed through the walls of the small intestine and stored in the liver until required in the red bone marrow for the normal process of development of the red blood cells.

Figure 225

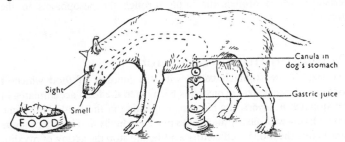

Illustration depicting the cephalic phase of gastric secretion.

SECRETION OF GASTRIC JUICE

There is always a small quantity of gastric juice present in the stomach, even when it contains no food. This is known as *fasting juice.*

There are three phases of secretion of gastric juice.

Cephalic Phase. This flow of juice occurs *before* food reaches the stomach and is due to reflex stimulation of the vagus nerves following the sight, smell or taste of food. In animals where the vagus nerves have been cut this phase of gastric secretion stops.

Gastric Phase. When stimulated by the presence of food the stomach produces a hormone called *gastrin* which passes directly into the circulating blood. This hormone circulating in the blood stimulates the glands in the stomach wall to produce more gastric juice. In this way the secretion of digestive juice is continued after the completion of the meal.

Intestinal Phase. When the partially digested contents of the stomach reach the duodenum, especially if they contain a considerable amount of fat, a hormone *enterogastrone* is produced which slows down the secretion of gastric juice and reduces gastric motility. By slowing the emptying rate of the stomach, the contents of the duodenum become more thoroughly mixed with bile and pancreatic juice.

FUNCTIONS OF THE STOMACH

1. The stomach acts as a temporary reservoir for food thus allowing the digestive juices time to act on the different food substances.

2. It produces *gastric juice* which begins the chemical digestion of proteins.

3. Muscular action includes the churning and onward movement of the stomach contents. When the contents of the pyloric end of the stomach have reached the required degree of acidity and liquefaction they are known as *chyme.* Chyme is passed through into the duodenum in small jets when the pyloric sphincter relaxes and the muscular walls of the stomach contract. This can happen only when the gastric contents have reached the required consistency.

4. Absorption takes place to a limited degree. Water, glucose, alcohol and some drugs are absorbed through the walls of the stomach into the venous circulation.

5. Although iron absorption takes place in the small intestine it is dissolved out of foods most effectively in the stomach in the presence of hydrochloric acid.

Digestion in the Small Intestine

When acid chyme passes into the small intestine it is mixed first with the *pancreatic juice* and *bile* then with *intestinal juice.* Although intestinal juice is secreted by the walls of the duodenum as well as the jejunum and the ileum its action is very limited in this area. It is the juice which *completes* the digestion of carbohydrates to monosaccharides, proteins to amino acids and fats to fatty acids and glycerol.

PANCREATIC JUICE

Pancreatic juice enters the duodenum at the ampulla of the bile duct and consists of:

water
mineral salts
enzymes; trypsinogen, chymotrypsinogen, amylase, lipase.

Acid chyme is rendered alkaline by the strong alkalinity of the pancreatic juice (pH 8·5). This alteration in reaction is necessary for the effective action of the pancreatic enzymes.

FUNCTIONS

Trypsinogen and *chymotrypsinogen* are inactive enzymes until they come in contact with *enterokinase* of the intestinal juice which converts them into *trypsin* and *chymotrypsin*. These enzymes convert peptones into peptides and polypeptides.

Amylase converts *all* polysaccharides (starches) not affected by ptyalin to disaccharides (sugars).

Lipase converts fats into fatty acids and glycerol. To aid the action of lipase *bile salts* emulsify the fats, breaking the fat down into smaller globules.

SECRETION

The secretion of pancreatic juice is stimulated by two hormones *secretin* and *pancreozymin* produced by the walls of the small intestine. The presence of acid chyme in the duodenum precipitates the production of these hormones.

BILE

Bile, secreted by the liver, passes down the *hepatic duct* and the *bile duct*, but is unable to enter the duodenum when the sphincter of Oddi is closed, therefore it passes along the *cystic duct* to the gall bladder where it is stored. When a meal has been taken the gall bladder contracts, the sphincter of Oddi relaxes and bile passes freely into the duodenum. The hormone, *cholecystokinin,* produced by the walls of the duodenum is believed to stimulate this activity. A more marked activity is noted if a fatty meal has been taken and chyme entering the duodenum contains a high proportion of fat.

COMPOSITION OF BILE

There is a slight difference between the composition of bile produced by the liver and that entering the duodenum. In the gall bladder water is absorbed and mucus is added by the goblet cells in the mucous membrane of the gall bladder, therefore the bile is concentrated and becomes a more viscid fluid.

The constituents of gall bladder bile are:
water
mineral salts
mucus
bile salts
bile pigments—bilirubin and biliverdin
cholesterol.

FUNCTIONS

1. The bile salts, *sodium taurocholate* and *sodium glycocholate*, are active in emulsifying fats in the duodenum.

2. Bile pigments, *bilirubin* and *biliverdin,* are the waste products of the breakdown of the red blood cells and bile is their route of excretion.

3. The presence of bile in the small intestine is necessary for the absorption of vitamin K and digested fats.

4. Bile colours and deodorises the faeces.

5. Bile has an aperient action.

INTESTINAL JUICE OR SUCCUS ENTERICUS

This is the digestive juice which completes the digestion of carbohydrates, proteins and fats. It is secreted by the specialised glands situated between the villi of the intestinal mucosa. It is alkaline in reaction (pH 8·0) and consists of:

> water
> mineral salts
> mucus
> enzymes—enterokinase which converts inactive trypsinogen and
> > chymotrypsinogen to active trypsin and chymotrypsin.
> > peptidases
> > lipase
> > sucrase, maltase and lactase
> > cholecystokinin—upper intestine only.

FUNCTIONS

Peptidases complete the digestion of protein by converting peptides and polypeptides to *amino acids.*

Lipase completes the digestion of fats to *fatty acids and glycerol.*

Sucrase, maltase and *lactase.* These enzymes complete the digestion of carbohydrates by converting disaccharides or sugars to *monosaccharides* of which glucose is the most commonly occurring.

SECRETION

Chyme entering the small intestine stimulates the mucous membrane to secrete the hormone *enterocrinin* which stimulates the intestinal glands to secrete intestinal juice. Another factor which may influence the flow of intestinal juice is the presence of nutritional materials in the small intestine. In this case there is reflex stimulation of secretion by distension and by direct contact between the food and the mucous membrane.

FUNCTIONS OF THE SMALL INTESTINE

1. Onward movement of its contents which is produced by peristalis and segmental and pendular movements.

2. Secretion of intestinal juice.

3. Completion of digestion of carbohydrates, proteins and fats.

4. Protection against infection by micro-organisms, which have

survived the bactericidal action of the hydrochloric acid in the stomach, by the solitary lymph nodes and aggregated glands.

Figure 226

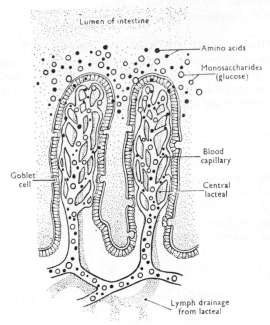

Diagram depicting the absorption of amino acids and glucose into the capillaries of the villi.

5. Absorption of nutritional materials. Carbohydrates, proteins and fats in their undigested or partly digested state cannot pass through the mucous membrane of the small intestine into the body; but in the form of glucose, amino acids, fatty acids and glycerol they can permeate through the columnar epithelial cells, which form the walls of the *villi,* into the *lacteals* and *blood capillaries* of the villi. *Glucose and amino acids* are absorbed into the *blood capillaries. Fatty acids and glycerol* are absorbed into the *lacteals* giving a milky appearance to the lymph which is now called *chyle.* Other nutritional materials such as vitamins, mineral salts and water are absorbed from the small intestine into the blood capillaries.

It is to be remembered that the surface area through which absorption can take place is vastly increased by the *circular folds* of mucous membrane and by the large number of *villi.* It is believed that the absorption area of the small intestine is about five times that of the skin surface of the body.

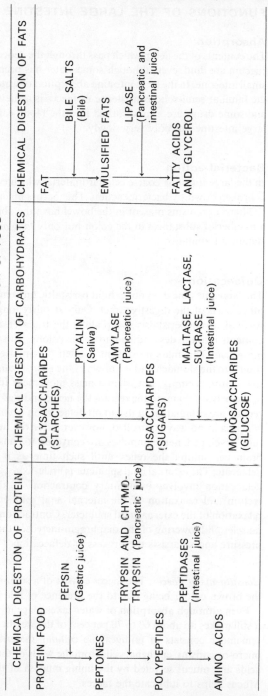

SUMMARY OF CHEMICAL DIGESTION OF FOOD

CHEMICAL DIGESTION OF PROTEIN	CHEMICAL DIGESTION OF CARBOHYDRATES	CHEMICAL DIGESTION OF FATS
PROTEIN FOOD	POLYSACCHARIDES (STARCHES)	FAT
↓ PEPSIN (Gastric juice)	↓ PTYALIN (Saliva)	↓ BILE SALTS (Bile)
PEPTONES	↓ AMYLASE (Pancreatic juice)	EMULSIFIED FATS
↓ TRYPSIN AND CHYMO-TRYPSIN (Pancreatic juice)	DISACCHARIDES (SUGARS)	↓ LIPASE (Pancreatic and intestinal juice)
POLYPEPTIDES	↓ MALTASE, LACTASE, SUCRASE (Intestinal juice)	FATTY ACIDS AND GLYCEROL
↓ PEPTIDASES (Intestinal juice)	MONOSACCHARIDES (GLUCOSE)	
AMINO ACIDS		

FUNCTIONS OF THE LARGE INTESTINE

Absorption

The contents of the ileum which pass through the ileocolic valve into the caecum are fluid, even although some water has been absorbed in the small intestine. In the large intestine absorption of water continues until the familiar semi-solid consistence of faeces is achieved. Mineral salts and some drugs are also absorbed into the blood capillaries from the large intestine but only very slowly.

Bacterial action

In the large intestine coarse cellular material which cannot be digested is broken down by bacterial action. This occurs mainly in the caecum.

Micro-organisms present in the bowel have the ability to synthesise a number of substances in the colon but only *folic acid* is absorbed in significant amounts.

Defaecation

The large intestine does not exhibit peristaltic movement as it is seen in other parts of the digestive tract. Only at fairly long intervals does a wave of strong peristalsis sweep along the transverse colon forcing its contents into the descending and pelvic colons. This is known as *mass movement*. This mass movement is often precipitated by the entry of food into the stomach and is known as the *gastro-colic reflex*. Normally the rectum is empty, but when a mass movement forces the contents of the pelvic colon into the rectum the nerve endings in the walls of the rectum are stimulated. In the infant defaecation occurs by reflex action which is in no way controlled, however, after the nervous system has fully developed, nerve impulses are conveyed to consciousness and the brain can inhibit the reflex until such time as it is convenient to defaecate. The external anal sphincter is under conscious control. Thus defaecation involves involuntary contraction of the muscle of the rectum and relaxation of the internal anal sphincter and voluntary relaxation of the external anal sphincter. Contraction of the abdominal muscles and lowering of the diaphragm increases the intra-abdominal pressure and so assists the process of defaecation.

Constituents of Faeces. The faeces consist of a semi-solid brown mass, the brown colour being due to the presence of bile pigments.

Even although absorption of water takes place in the large intestine it still makes up about 60 to 70 per cent of the weight of the faeces. The remainder consists of undigestible cellular material, dead and live micro-organisms, epithelial cells from the walls of the tract, some fatty acids and mucus secreted by the lining mucosa of the large intestine. Mucus helps to lubricate the faeces.

Metabolism

Metabolism is the name given to the sum total of all the chemical changes which take place within the body. Reactions which involve the breakdown or decomposition of a substance are called *katabolic* and those involving the building up or synthesis of new material are called *anabolic*. An example of katabolism would be the breakdown of the red blood cells and an example of anabolism would be the building up of absorbed amino acids into complex proteins within the body, such as, liver tissue.

Figure 228

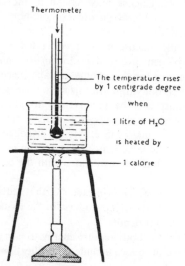

Drawing showing the meaning of the word Calorie.

To carry out these chemical changes within the body energy is necessary. When energy is used heat is produced and it is possible to measure the rate of heat production. This is known as the *metabolic rate* or the rate at which metabolic changes take place. The large *Calorie* or kilocalorie is the unit of heat used for such measurement, and is defined as the amount of heat required to raise the temperature of one litre of water through one centigrade degree.

The basal metabolic rate is the standard used in measurement. The basal metabolic rate is the rate at which metabolic changes take place during *resting and fasting*. It is determined by calculation from the amount of *carbon dioxide excreted* or the amount of *oxygen used* in a measured time.

The source of energy to carry out these metabolic processes is the nutritional materials, and the relative energy producing ability of each of these is noteworthy.

One gramme of carbohydrate provides 4·1 Calories.
One gramme of protein provides 4·1 Calories.
One gramme of fat provides 9·3 Calories.

METABOLISM OF CARBOHYDRATE

When digested, carbohydrate in the form of monosaccharides mainly glucose, is absorbed into the blood capillaries of the villi of the small intestine. It is transported by the portal circulation to the liver, where it is dealt with in one of several ways.

1. Glucose may be used to provide the energy necessary for the considerable metabolic activity which takes place in the liver.

2. Some of the glucose may remain in the circulating blood to maintain the normal blood glucose level of about 120 mg per cent (per 100 ml of blood).

3. Some of the glucose may be converted to the insoluble polysaccharide, glycogen, in the liver and in the muscles. *Insulin* is the hormone necessary for this change to take place. Glycogen forms a useful store of carbohydrate. In this form it cannot be used to provide energy, but must first be reconverted to glucose. Liver glycogen constitutes a store for the supply of glucose for liver activity and to maintain the blood glucose level. Muscle glycogen provides the glucose requirement of muscle activity. *Adrenalin, thyroxin,* and *glucagon* are three hormones which are associated with the convertion of glycogen to glucose.

4. Carbohydrate which is in excess of that required to maintain the blood glucose level and glycogen level in the tissues is converted to fat and stored in the fat depots.

All the cells of the body require energy to carry out their metabolic processes. These processes include multiplication of cells for replace-

Figure 229

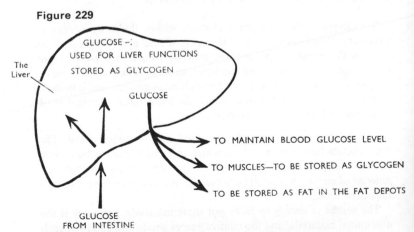

Schematic drawing showing the utilisation of glucose in the body.

ment of worn out cells, contraction of muscle fibres, secretion by the cells of a gland, for example, the pancreas. Glucose and fat are the main sources of the energy required by the body.

Some energy can be provided by glucose in the absence of oxygen. This anaerobic process is uneconomical and very limited. This is the energy used in a sudden spurt of activity over a very short period of time, for example, the man who runs 100 yards in 10 seconds could not take in enough oxygen in that time to provide energy by the complete oxidation of glucose, so he has to depend on the anaerobic process of energy release. One of the end products of this process is lactic acid, and if it accumulates in excess in the muscles it causes the pain which is associated with unaccustomed exercise.

Complete oxidation of glucose requires an adequate supply of oxygen. This is the release of energy for more prolonged but less arduous activity, for example, the man who runs a mile in four minutes depends upon aerobic oxidation. The energy release takes place more slowly and is balanced by oxygen intake. The pace of the mile runner is considerably less than his sprinting counterpart but his period of activity is longer. Complete oxidation of carbohydrate in the body results in the production of the end products carbon dioxide and water.

THE FATE OF THE END PRODUCTS OF CARBOHYDRATE METABOLISM

1. *Lactic Acid.* Some of the lactic acid produced by anaerobic metabolism of glucose may be oxidised in the tissues to carbon dioxide and water but first it must be changed to pyruvic acid. If complete oxidation does not take place lactic acid passes to the liver in the circulating blood where it is converted to glucose and may then take any of the pathways open to glucose.

2. *Carbon Dioxide.* This is excreted from the body as a gas by the lungs. This has already been described in external respiration.

3. *Water.* The water of metabolism is added to the considerable amount of water already present in the body.

METABOLISM OF PROTEIN

Protein foods which are taken as part of the diet consist of a number of amino acids. About *20 amino acids* have been named and about 10 of these are described as *essential* because they can not be synthesised in the body, the remainder are described as *non-essential* amino acids because they can be synthesised by many tissues. The enzymes involved in this process are called *transaminases.* Digestion breaks down the protein of the diet to its constituent amino acids in preparation for absorption into the blood capillaries of the villi. In the portal circulation amino acids are transported to the liver then into the general circulation, thus making them available to all the cells and tissues of the body. Different cells choose from those available the particular amino acids required for building or repairing that specific type of tissue. It is as though a building has been demolished and the bricks transported to a new site. Some of these bricks may be used to build one type of house and some another type and there may be some which cannot be used and are, therefore, discarded. To explain the analogy the demolished building should be regarded as the protein materials in the diet, the amino acids as the bricks and the different tissues of the body as the new buildings.

Figure 230

Drawing depicting the meaning of the words katabolism and anabolism.

Amino acids which are not required for building and repairing body tissues are broken down into two parts in the liver. First, the *nitrogenous part* which is changed into *urea* and excreted in the urine and second, a part which is used as a fuel to provide heat and energy and will finally be oxidised to carbon dioxide and water. This whole process is called the *deamination of protein*.

The body is unable to *store* nitrogenous materials in the form of amino acids therefore a regular supply, through the diet, is essential.

METABOLISM OF FAT

Fats which have been digested and absorbed into the *lacteals* are transported via the receptaculum chyli and the thoracic duct to the bloodstream and so, by a circuitous route, to the liver. In the liver the fatty acids and glycerol are reorganised and recombined to form human fat and as such is stored in the fat depots of the body. These include subcutaneous fat, fat supporting the kidneys, and between the layers of the greater omentum. Before this stored fat can be used to provide energy it has to be transported back to the liver, undergo further change called *desaturation*, then it is passed into the circulation in a form which can gain entry to all cells and be oxidised to carbon dioxide and water.

Ketone bodies are produced during the process of oxidation of fats and are always present in the blood in very small amounts. They are excreted in the urine and in the expired air as acetone. When there has been an insufficient intake of carbohydrate foods in the diet, fat is used up in excessive quantities to provide energy and heat, thus a state of ketosis arises due to the increase in the amount of ketone bodies in the blood.

Figure 231

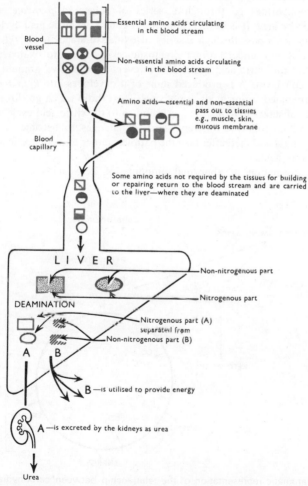

Essential amino acids circulating in the blood stream

Blood vessel

Non-essential amino acids circulating in the blood stream

Amino acids—essential and non-essential pass out to tissues e.g., muscle, skin, mucous membrane

Blood capillary

Some amino acids not required by the tissues for building or repairing return to the blood stream and are carried to the liver—where they are deaminated

L I V E R

Non-nitrogenous part

Nitrogenous part

DEAMINATION

Nitrogenous part (A) separated from Non-nitrogenous part (B)

A   B

B —is utilised to provide energy

A —is excreted by the kidneys as urea

Urea

Scheme showing the fate of amino acids in the body.

RELATIONSHIPS BETWEEN CARBOHYDRATES, FATTY ACIDS AND DEAMINATED AMINO ACIDS AS ENERGY RELEASING SUBSTANCES

The degradation of carbohydrates, fatty acids and deaminated amino acids occurs inside the cells releasing *energy* and forming the waste products *carbon dioxide* and *water*. The katabolism of these molecules occurs in a series of steps, a little energy being released at each stage. Up to a certain point each nutrient passes through a series of separate and distinct stages but thereafter, they all follow a common pathway of degradation. This final common pathway is called the *citric acid cycle* or Kreb cycle.

Figure 232 provides a diagrammatic representation of the processes involved and only a few of the many steps are shown.

Carbohydrates go through a series of stages to *pyruvate* and *oxaloacetic acid*. It is in this form that it enters the citric acid cycle.

Fatty acids pass through a series of oxidative stages to *acetyl co-enzyme A* and, under normal circumstances, progress to oxaloacetic acid and the citric acid cycle. If, however, an excessive amount of acetyl co-enzyme A is produced some of it developes into *keto acids*.

Deaminated amino acids are of two types: those which go through a series of stages to *oxaloacetic acid* and so to the citric acid cycle and those which follow a different series of changes to become *acetyl co-enzyme A* and thereafter take the pathway either to oxaloacetic acid or to keto acids.

The formation of abnormal amounts of keto acids occurs in

Figure 232

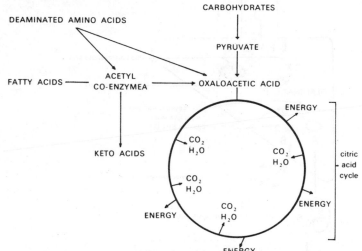

Diagrammatic representation of the relationship between carbohydrates, fatty acids and deaminated amino acids as energy releasing substances.

starvation and in diabetes mellitus when excessive amounts of fat and amino acids are used to provide energy, that is, when acetyl co-enzyme A is produced more rapidly than it can be used in the citric acid cycle. In both these examples there is an insufficiency of carbohydrate inside the cells. In diabetes this is due to a shortage in the supply of the hormone *insulin* which facilitates the transportation of carbohydrate from the extracellular fluid across the cell membrane. Excess keto acids are excreted in the urine and in the expired air as acetone.

FUNCTIONS OF THE LIVER

The liver is an extremely active organ. Some of its functions have already been described therefore they will only be mentioned here.

1. *Converts glucose to glycogen* in the presence of *insulin*, and changes the liver glycogen back to glucose as it is required.

2. *Deamination of protein.*
   (a) The removal of the nitrogenous portion from the amino acids which are not required for building and repairing body tissues, and the formation of *urea* from this nitrogenous portion.
   (b) The breakdown of the nucleo-protein of the worn-out cells of the body to form *uric acid* which is excreted in the urine.
   (c) The formation of urea from the protoplasm of worn-out cells.

3. *Desaturation of fats*, that is, converting stored fat to a form in which it can be used by the tissues to provide energy.

4. *Heat production.* The liver uses a considerable amount of energy, has a high metabolic rate and produces a great deal of heat. It is the main heat producing organ of the body.

5. *Secretion of bile.* The constituents of bile are not all present in the blood as such, but the liver cells have the ability to form these substances from the mixed venous and arterial blood in the sinusoids, for example, bile salts and cholesterol.

6. *Storage of the anti-anaemic factor.* The anti-anaemic factor is absorbed through the small intestine and is transported in the portal circulation to the liver where it is stored until required in the red bone marrow for the maturation of the red blood cells.

7. *Stores iron* derived from:
   (a) the diet.
   (b) the break down of worn-out red blood cells in the spleen.

8. *Stores vitamins A, D, E and K* which have been taken into the body in the diet.

9. *Synthesis of vitamin A.* This vitamin can be formed in the liver from carotene, the provitamin found in certain plants, for example, in carrots and green leaves of vegetables.

10. *Forms the plasma proteins, serum albumin and serum globulin,* from the available amino acids.

11. *Forms prothrombin and fibrinogen* from the available amino acids. These proteins circulate in the blood and are essential for the coagulation of blood.

SUMMARY OF DIGESTION, ABSORPTION AND UTILISATION OF CARBOHYDRATES, PROTEINS AND FATS

CARBOHYDRATES

Digestion

Organ	Digestive juice	Enzyme and Action
Mouth	Saliva	Ptyalin converts cooked starches to *maltose*
Stomach	Gastric juice	Hydrochloric acid stops the action of salivary ptyalin
Small Intestine	Pancreatic juice	Amylase converts all starches to *disaccharides* (sugars)
Small Intestine	Intestinal juice	Sucrase⎫ Convert all sugars to Maltase⎬ *monosaccharides,* Lactase⎭ mainly *glucose*

Absorption

Glucose is absorbed into the capillaries of the villi and transported in the portal circulation to the liver.

Utilisation

1. For liver metabolism.
2. To maintain a constant blood glucose level so that all body tissues have a constant supply.
3. Some of the excess is converted to glycogen in the presence of insulin and stored in the liver and in the muscles.
4. Any remaining glucose is converted into fat and stored in the fat depots.
5. Glucose is used in the body to provide energy and heat. Oxygen is necessary for its complete break down and the waste products left are carbon dioxide and water.

PROTEIN

Digestion

Organ	Digestive Juice	Enzyme and Action
Mouth	Saliva	No action
Stomach	Gastric juice	Hydrochloric acid converts pepsinogen to *pepsin* Pepsin converts all proteins, to *peptone* Rennin converts soluble caseinogen to insoluble *casein*

Small intestine	Pancreatic juice	Enterokinase of intestinal juice converts trypsinogen and chymotrypsinogen to *trypsin* and *chymotrypsin* which convert peptones to *peptides* and *polypeptides*
Small Intestine	Intestinal juice	Peptidases convert peptides and polypeptides to *amino acid*

Absorption

Amino acids are absorbed into the capillaries of the villi and transported in the portal circulation to the liver.

Utilisation

1. In the liver to form serum albumin, serum globulin, prothrombin and fibrinogen.

2. Used in various combinations by cells of the body for cell multiplication, cell repair and the production of secretions, e.g. hormones, enzymes.

3. Amino acids which are not required are deaminated in the liver. The nitrogenous part is converted into urea and excreted in the urine. The remaining part is used to provide energy and heat or deposited as fat in the fat depots.

FATS

Digestion

Organ	Digestive Juice	Enzyme and Action
Mouth	Saliva	No action
Stomach	Gastric juice	No action
Small Intestine	Bile	Bile salts emulsify fats
Small Intestine	Pancreatic juice	Lipase converts fats to *fatty acids* and *glycerol*
Small Intestine	Intestinal juice	Lipase completes the digestion of fats to *fatty acids* and *glycerol*

Absorption

Fatty acids and glycerol are absorbed into the lacteals of the villi and are transported via the receptaculum chyli and the thoracic duct to the left subclavian vein. In this way they are transported by the circulating blood to the liver where fatty acids and glycerol are reorganised and recombined.

Utilisation

1. Utilised in the presence of oxygen to provide energy and heat, the waste products carbon dioxide and water being produced.

2. Stored in the fat depots.

When depot fat is required for oxidation it must first be desaturated by the liver.

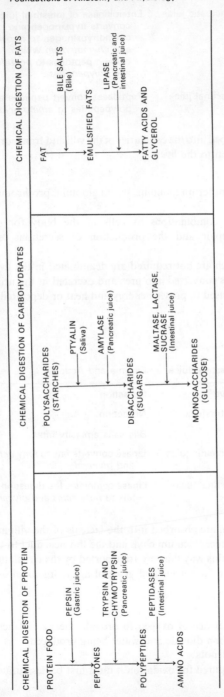

SUMMARY OF CHEMICAL DIGESTION OF FOOD

CHEMICAL DIGESTION OF PROTEIN

PROTEIN FOOD
↓ PEPSIN (Gastric juice)
PEPTONES
↓ TRYPSIN AND CHYMOTRYPSIN (Pancreatic juice)
POLYPEPTIDES
↓ PEPTIDASES (Intestinal juice)
AMINO ACIDS

CHEMICAL DIGESTION OF CARBOHYDRATES

POLYSACCHARIDES (STARCHES)
↓ PTYALIN (Saliva)
↓ AMYLASE (Pancreatic juice)
DISACCHARIDES (SUGARS)
↓ MALTASE, LACTASE, SUCRASE (Intestinal juice)
MONOSACCHARIDES (GLUCOSE)

CHEMICAL DIGESTION OF FATS

FAT
↓ BILE SALTS (Bile)
EMULSIFIED FATS
↓ LIPASE (Pancreatic and intestinal juice)
FATTY ACIDS AND GLYCEROL

12. The Urinary System

The urinary system is one of the excretory systems of the body.

The organs of the urinary system consist of:

2 *kidneys* which form and secrete urine.

2 *ureters* which convey the urine from the kidneys to the urinary bladder.

1 *urinary bladder* where urine is temporarily stored.

1 *urethra* through which the urine is discharged from the urinary bladder to the exterior.

Figure 233

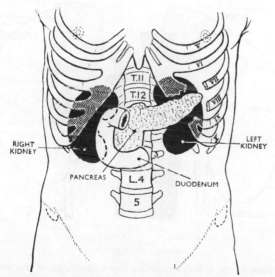

Position of the kidneys in association with the pancreas, duodenum, ribs and vertebral column.

The Kidneys

The kidneys are situated in the posterior part of the abdomen, one on each side of the vertebral column, behind the peritoneum.

They extend from approximately the twelfth thoracic vertebra above to the third lumbar vertebra below. The right kidney is usually slightly lower than the left, probably due to the considerable amount of space occupied by the liver. The left kidney is slightly longer and narrower than the right and lies nearer to the median plane.

The kidneys are described as bean-shaped organs and are approximately 12 cm (4½ inches) long, 7 cm (2 to 3 inches) wide and 4 cm (1 to 1½ inches) thick. They are embedded and held in position by a mass of adipose tissue termed the *renal fat*. Surrounding the kidney and the renal fat is a sheath of fibro-elastic tissue known as the *renal fascia*.

ORGANS IN ASSOCIATION WITH THE KIDNEYS

As the kidneys lie on either side of the vertebral column the organs in association with each kidney differ.

The right kidney

Superiorly. The right *supra-renal gland* lies on the upper border of the kidney like a dunce's cap.

Figure 234

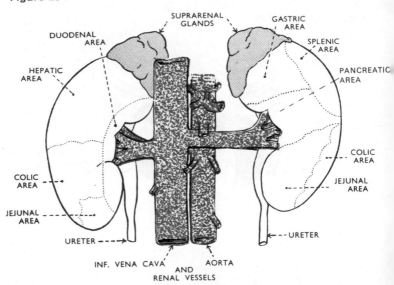

Anterior view of the kidneys showing the impressions of the associated organs.

Anteriorly. The right lobe of the *liver*, the *duodenum* and the *right colic flexure* of the large intestine.

Posteriorly. The diaphragm, and the muscles of the posterior abdominal wall.

The left kidney

Superiorly. The left *supra-renal gland* lies on its upper border.

Anteriorly. The *spleen*, the *stomach*, the *pancreas*, the *jejunum* and the *left colic flexure* of the large intestine.

Posteriorly. The *diaphragm* and the muscles of the posterior abdominal wall.

Figure 235

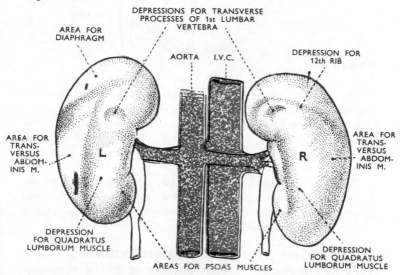

Posterior view of the kidneys showing the impressions of the associated organs.

THE GROSS STRUCTURE OF THE KIDNEY

The kidneys are enclosed in a *capsule of fibrous tissue* which can be stripped off quite easily. Underlying the capsule is the external part of the kidney which is reddish brown in colour and is known as the *cortex*. Lying internally is the *medulla*. The medullary substance is made up of pale conical striations known as the *renal pyramids*. Medial to the medulla is a concave fissure which is known as the *hilum* of the kidney. At the hilum the *renal artery* and *renal nerves* enter the kidney and the *renal vein* leaves the kidney. It is here that the upper expanded part of the ureter, known as the *pelvis* leaves the kidney.

Figure 236

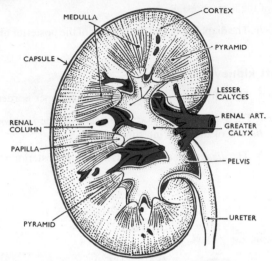

A kidney.

The hilum leads into a central cavity known as the *renal sinus* which contains the renal blood vessels and the pelvis of the ureter.

Within the renal sinus the pelvis of the ureter divides into a number of branches which are named the *calyces*. The apices of the renal pyramids converge towards the calyces where they form prominent *papillae* which project into them.

The cortical substance of the kidney has a smooth texture, but the medulla has a striated appearance due to the presence of the renal pyramids. The cortex extends inwards between each pyramid forming areas known as the *renal columns*.

MICROSCOPIC STRUCTURE OF THE KIDNEYS

The kidney substance is composed of a large number of microscopic structures known as *nephrons* and *collecting tubules*. The nephron is described as the *functional unit* of the kidney and it is estimated that there are approximately one million in each kidney.

The nephron

The nephron consists of a tubule which is closed at one end. The other end opens into a *collecting tubule*. The closed, or blind end, is indented to form a cup-shaped structure called the *glomerular capsule* which almost completely encloses a network of arterial capillaries called the *glomerulus*. Continuing from the glomerular capsule the remainder of the nephron is described in three parts; the *proximal convoluted tubule*, the *loop of Henle* and the *distal convoluted tubule* which leads into a collecting tubule.

Figure 237

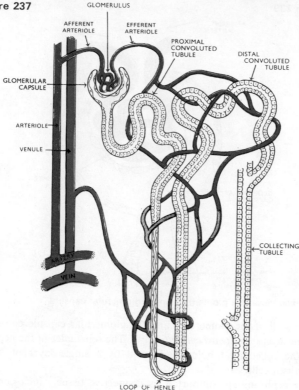

The nephron and its capillary network.

The renal artery after entering the kidney at the hilum, divides into small arteries and arterioles. In the cortex one arteriole, the *afferent arteriole*, enters each glomerular capsule where it divides to form the glomerulus. The blood vessel leading away from the glomerulus is called the *efferent arteriole* which breaks up into a second capillary network to supply oxygen and nutritional materials to the remainder of the nephron. Venous blood drained away from this capillary bed eventually leaves the kidney in the renal vein which empties into the inferior vena cava.

Figure 238

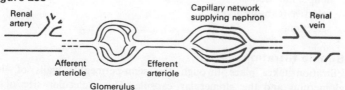

Figure 239

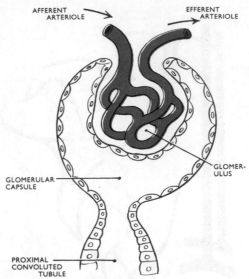

Magnified view of a glomerulus and a glomerular capsule.

The wall of the glomerulus and the glomerular capsule consist of a single layer of *flattened epithelial cells*. The remainder of the nephron and the collecting tubule are formed by a single layer of highly specialised cells.

The nephrons are supported by connective tissue which contains sympathetic and parasympathetic nerves.

FUNCTIONS OF THE KIDNEY
The kidneys are all important to the health of the individual. Their function is to form urine and pass the urine to the ureters and the urinary bladder for excretion. In doing this, vital functions in relation to the maintenance of fluid and electrolyte balance and the disposal of waste material from the body are carried out.

FORMATION OF URINE
Urine is formed by the nephrons of the kidney and the process occurs in three phases:

simple filtration
selective reabsorption
secretion.

Simple filtration
Filtration takes place through the semi-permeable walls of the glomerulus and the glomerular capsule. The molecular size of a substance present in the blood determines whether it is possible for it to

pass through the filter. The main factor which assists filtration is the difference between the blood pressure in the glomerulus and the filtrate pressure in the glomerular capsule. Because the calibre of the efferent arteriole is less than that of the afferent arteriole, a capillary hydrostatic pressure of about 90 mmHg builds up in the glomerulus. This pressure is opposed by the osmotic pressure of the blood, about 25 mmHg, and by filtrate hydrostatic pressure of about 15 mmHg inside the glomerular capsule.

The net filtration pressure is, therefore, $90 - 25 - 15 = 50$ mmHg.

Substances from the blood of a low molecular weight which filter through into the glomerular capsule include water, salts, amino acids, fatty acids, glucose, urea, uric acid, creatinine, hormones and toxins. The contents of the blood, such as the red blood corpuscles, white blood cells and plasma proteins which are of high molecular weight *do not* leave the glomerulus as they are unable to pass through this semi-permeable membrane.

Selective reabsorption

Selective reabsorption is the process by which the composition and volume of the glomerular filtrate is altered during its passage through the convoluted tubules, the loop of Henle and the collecting tubule. The general purposes of this process are the reabsorption of those constituents of the filtrate which are essential to the body, the maintenance of fluid and electrolyte balance and the maintenance of the alkalinity of the blood (pH 7·4).

Some constituents of the glomerular filtrate do not appear in the *urine* under normal circumstances, for example, glucose, amino acids, keto acids, Vitamin C and some salts. These substances are described as having *high threshold values*. However, the capacity of the tubules to reabsorb these substances is not unlimited, for example, the normal blood glucose level is about 120 mg per cent and at this level all the glucose in the filtrate is reabsorbed, but if the blood glucose level rises above 180 mg per cent, as it may in diabetes mellitus, all of it can not be reabsorbed and some appears in the urine.

Some substances such as creatinine and sulphates are not reabsorbed and are described as *non-threshold substances*.

Between these extremes there are some substances which are reabsorbed in varying amounts according to the conditions within the body. This is the type of reabsorption associated with the maintenance of the alkalinity of the blood and the fluid and electrolyte balance of the body.

Secretion

Filtration occurs as the blood flows through the glomerulus. Non-threshold substances and foreign materials, such as drugs, may not be cleared from the blood by filtration because it does not remain for a sufficient length of time in the glomerulus. Such substances are cleared from the blood by secretion *into* the convoluted tubules and passed from the body in the urine.

COMPOSITION OF URINE

Water, 96 per cent

Urea, 2 per cent

Uric acid

Creatinine

Phosphates   2 per cent

Sulphates

Oxylates

Chlorides

Urine is amber in colour, has a specific gravity of 1020–1030, and is acid in reaction. A healthy adult passes approximately 2 to 3 pints (1–1½ litres) in 24 hours.

WATER BALANCE

Water is taken into the body through the alimentary tract and a small amount formed by the metabolic processes is added to the total body water. The excretion of water occurs in saturated expired air, as a constituent of the faeces, through the skin as sweat and in the urine. The amount lost in expired air and in the faeces is fairly constant and the amount of sweat produced is associated with the maintenance of normal body temperature. The balance between intake and output is maintained, therefore, by the kidneys. The minimum urinary output which is consistent with the essential removal of waste materials from the body by this route is about 500 ml per day. The amount produced in excess of this is controlled mainly by the *antidiuretic hormone* (ADH) which is released into the blood by the posterior lobe of the pituitary gland. There is a close link between the hypothalamus in the brain and the posterior lobe of the pituitary gland and it has been shown that there are cells in the hypothalamus which are sensitive to changes in the osmotic pressure of the blood. These cells are called *osmoreceptors*. The link between the hypothalamus and the posterior lobe of the pituitary gland is provided by the circulating blood and by nerve fibres.

If the osmotic pressure of the blood is raised the osmo-receptors in the hypothalamus are stimulated and there is an increase in the output of antidiuretic hormone by the pituitary gland. Conversely, if the osmotic pressure of the blood is reduced the amount of antidiuretic hormone released into the blood is reduced.

The antidiuretic hormone acts on the tubules of the kidneys by increasing the permeability of the distal and collecting tubules and consequently the amount of water reabsorbed and returned to the circulating blood. If there is an increase in the concentration of dissolved substances in the blood there is an increase in the amount of

ADH produced, resulting in an increase in the reabsorption of water which dilutes the blood. The effect of diluting the blood is to reduce the amount of ADH produced. This cyclic effect maintains the concentration of the blood within narrow limits.

Figure 240

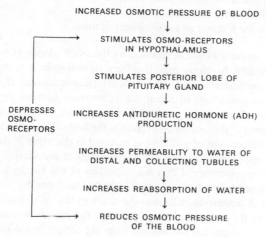

Feed-back mechanism which controls ADH output.

Although the antidiuretic hormone is the most important means by which the water balance of the body is maintained it is not the only one. If there is an excessive amount of any dissolved substance in the blood which must be excreted through the kidneys an excess of water is excreted with it. An example of this occurs in untreated diabetes mellitus. Because of the high concentration of glucose in the blood large amounts are excreted in the urine accompanied by an excess of water. Also, if excessive amounts of electrolytes have to be excreted from the body the amount of water required for this purpose is increased. This may lead to dehydration in spite of increased production of ADH but it is usually accompanied by acute thirst and increased water intake. The classical description of his condition by the patient with diabetes mellitus before treatment, includes an account of acute thirst and the passage of unusually large amounts of urine.

ELECTROLYTE BALANCE

Changes in the concentration of electrolytes in the body may be due, either to changes in the amounts of water or to the amounts of electrolytes. In its physiological functions the body has certain mechanisms which, in the healthy individual, maintain the balance between water and electrolyte concentration.

Sodium is the most common of the cations to exist in the extracellular fluid of the body. It is a constituent of almost all the foods which are taken in the diet and is often added to food as common salt during cooking. In the healthy individual, therefore, there is little danger of the intake of sodium being inadequate.

Loss of sodium from the body occurs by two main routes:
through the skin as a constituent of sweat
through the kidneys as a constituent of urine.

As the amount of sweat secreted by the sweat glands in the skin is closely related to physiological efforts to maintain a normal body temperature, sodium loss by this route is largely incidental. If, however, there is excessive loss of sweat the concurrent sodium depletion is significant. The kidneys, on the other hand have an active role in the maintenance of sodium concentration in the body. The substance which controls the amount of sodium ion excreted in the urine is *aldosterone,* a hormone produced by the cells of the cortex of the adrenal gland.

It will be remembered that the nephrons of the kidney have three main functions, filtration, selective reabsorption from the filtrate, and secretion. Aldosterone influences the reabsorption of sodium from the filtrate. As the amount of aldosterone in the blood increases so the amount of sodium ion reabsorbed from the filtrate in the convoluted tubules into the blood stream increases, thus maintaining the normal cation concentration in the extracellular fluid. This mechanism has an effect on the potassium level in the extracellular fluid and indirectly on the intracellular potassium. When the amount of sodium reabsorbed is increased the amount of potassium excreted is increased and vice versa. A tumour, therefore, of the adrenal cortex which results in the excessive production of aldosterone will cause the retention of a large amount of sodium and an excessive loss of potassium. On the other hand, adrenalectomy would result in sodium depletion and potassium retention, if maintenance doses of the hormone aldosterone were not given to the patient.

The secretion of aldosterone by the adrenal cortex is stimulated by the adrenocorticotropic hormone (ACTH) produced by the anterior lobe of the pituitary gland, a fall in the concentration of sodium in the blood and by *renin* produced in the kidneys.

When the blood pressure in the afferent arteriole falls renin is produced by the cells surrounding the arteriole. Renin acts on serum globulin to produce *angiotensin* which stimulates the adrenal cortex to produce aldosterone. The resultant reabsorption of sodium, accompanied by water, increases the blood volume which raises the blood pressure and reduces the renin output. Angiotensin has a powerful vaso-constrictor effect on the blood vessels generally.

Sodium and potassium are in high concentration in digestive juices, sodium in gastric juice and potassium in pancreatic and intestinal juice.

Figure 241

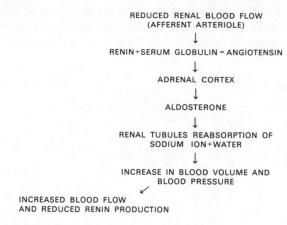

REDUCED RENAL BLOOD FLOW
(AFFERENT ARTERIOLE)
↓
RENIN+SERUM GLOBULIN = ANGIOTENSIN
↓
ADRENAL CORTEX
↓
ALDOSTERONE
↓
RENAL TUBULES REABSORPTION OF
SODIUM ION+WATER
↓
INCREASE IN BLOOD VOLUME AND
BLOOD PRESSURE
↙
INCREASED BLOOD FLOW
AND REDUCED RENIN PRODUCTION

Summary of the relationship between renal blood pressure, renin and aldosterone production.

Normally these ions are reabsorbed by the colon but in acute and prolonged diarrhoea they may be excreted in large quantities with resultant electrolyte imbalance.

The control of the cation concentration of the body is two-fold. Initially by the buffering mechanisms of the body and then by the disposal of cations by the lungs and kidneys. If these systems do not successfully control the cation concentration, acidosis occurs. There are three main sources of cations which may cause acidosis.

Metabolic acidosis
When metabolism is normal, the buffering systems of the body maintain the pH of the body fluids within normal range. Pathological production of cations is seen in diabetic acidosis where fat metabolism is incomplete and intermediate products of fat metabolism in the form of keto acids are present in the blood in excessive quantities.

Renal acidosis
Renal acidosis occurs when the kidneys fail to excrete phosphate and sulphate and fail to form ammonia.

Respiratory acidosis
Respiratory acidosis occurs when the lungs fail to excrete sufficient amounts of carbon dioxide. This may be due to:
changes in the lungs which prevent normal gas exchange
depression of the respiratory centre by drugs
disease or paralysis of the muscles of respiration.

The Ureters

The ureters are the two tubes which convey the urine from the kidneys to the urinary bladder. Each tube measures approximately 25 to 30 cm (10 to 12 inches) in length, and its diameter is approximately that of a goose quill.

Figure 242

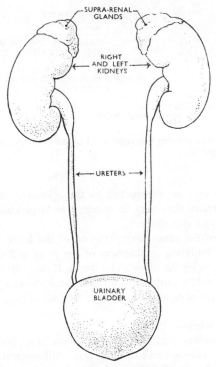

The kidneys in association with the suprarenal glands, ureters and urinary bladder.

The ureter commences as a funnel-shaped structure termed the *pelvis of the kidney*. It passes downwards through the abdominal cavity, behind the peritoneum and in front of the psoas muscle into the pelvic cavity, and opens into the posterior aspect of the base of the urinary bladder. The ureter passes obliquely through the bladder wall. When the bladder contracts the walls of the ureter are pressed together preventing urine from being forced back up the ureters.

STRUCTURE OF THE URETERS

The ureters consist of three layers of tissue.

1. An outer coat of *fibrous tissue*. This is continuous with the fibrous capsule of the kidney.

Figure 243

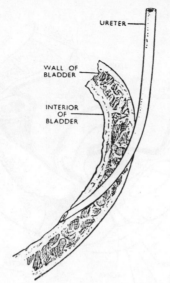

Diagram showing oblique entry of ureter into bladder.

2. A middle *muscular layer* consisting of smooth muscle fibres in two layers. An outer layer of longitudinal and an inner layer of circular muscle fibres.

3. An inner lining of *mucous membrane* consisting of transitional epithelium *on a basement membrane*.

FUNCTION OF THE URETERS
The ureters propel the urine from the kidney into the bladder by peristaltic contraction of the muscle layer.

The Urinary Bladder
The urinary bladder is described as a sac which acts as a reservoir for urine. It lies in the pelvic cavity, but its size and position vary depending on the amount of urine it contains. When grossly distended the bladder rises into the abdominal cavity.

ORGANS IN ASSOCIATION WITH THE BLADDER
The organs in association with the bladder differ in the male and the female.

In the female
Anteriorly. The symphysis pubis.
Posteriorly. The uterus.
Superiorly. The small intestine.
Inferiorly. The urethra and the muscles forming the pelvic floor.

Figure 244

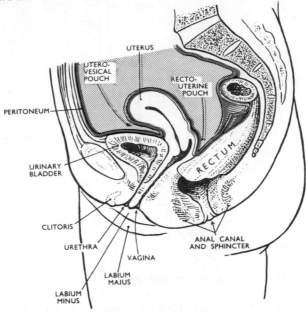

(A) In the female.

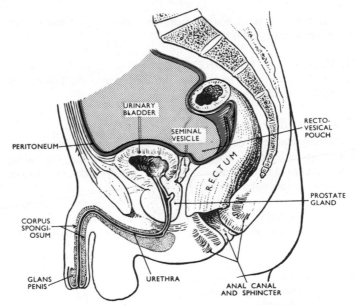

(B) In the male.

Diagram of the organs in association with the urinary bladder.

In the male

Anteriorly. The symphysis pubis.
Posteriorly. The rectum and seminal vesicles.
Superiorly. The small intestine.
Inferiorly. The urethra and prostate gland.

STRUCTURE OF THE BLADDER

The bladder is roughly pear shaped, but becomes more oval in shape
as it fills with urine. It is described as having anterior, superior and
posterior surfaces. The posterior surface of the bladder is known as the
base. The bladder opens into the urethra at its lowest point, *the neck*.

Figure 245

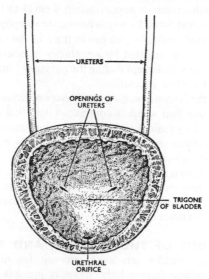

URETERS

OPENINGS OF
URETERS

TRIGONE
OF BLADDER

URETHRAL
ORIFICE

Position of the trigone of the bladder.

The bladder is composed of four layers of tissue.

1. *Peritoneum*. This serous membrane covers only the superior surface
of the bladder from which it is reflected upwards to become the parietal
peritoneum of the abdominal cavity. Posteriorly it is reflected on to
the uterus in the female and the rectum in the male.

2. *Muscle*. This layer is composed of longitudinal and circular muscle
fibres.

3. *Submucous*. This coat joins the inner lining with the muscular layer
and is made up of areolar tissue containing blood vessels, lymphatics
and sympathetic and parasympathetic nerves.

4. *Mucous Membrane*. This is the inner lining and is composed of
transitional epithelium. When the bladder is empty or contracted the
inner lining appears in folds. These folds gradually disappear as the
bladder distends with urine.

On examining the interior of the bladder three orifices can be seen. The upper two orifices on the posterior wall are the openings of the two ureters. The inferior orifice is the point of origin of the urethra. The three orifices form a triangle which is described as the *trigone* of the bladder. Where the urethra commences there is a *sphincter muscle* which controls the passage of urine.

THE URETHRA

The urethra is a canal which extends from the neck of the bladder to the exterior. The length of the urethra differs in the male and the female. The male urethra is associated with the reproductive system, and is described in Chapter 17.

The female urethra is approximately 4 cm (1 to 1½ inches) in length. It runs downwards and forwards behind the symphysis pubis and opens at the *external urethral orifice* just in front of the vagina. The external urethral orifice is guarded by a sphincter muscle which is under the control of the will. Except during the passage of urine the walls of the urethra are in close apposition.

The urethra is composed of three layers of tissue.

1. *A muscular coat* which is continuous with that of the bladder and consists of longitudinal and circular smooth muscle fibres. Near the external urethral orifice the smooth muscle is replaced by striated muscle fibres which are under voluntary control.

2. *A thin spongy coat* containing large numbers of blood vessels.

3. *A lining of mucous membrane* which is continuous with that of the bladder in the upper part of the urethra, the lower part consists of stratified squamous epithelium.

FUNCTIONS OF THE BLADDER AND MICTURITION

The urinary bladder acts as a reservoir for urine. As the urine gradually collects there is little change of pressure within the bladder for some time. The bladder as a whole adapts itself to the increased volume. When approximately 200 to 300 ml (7 to 10 oz.) of urine have accumulated the pressure stimulates the autonomic nerve endings within the bladder wall. In the infant micturition occurs by reflex action which is in no way controlled. However, after the nervous system has fully developed, nerve impulses are conveyed to consciousness, and the brain can inhibit the reflex for a limited period of time until it is convenient to micturate Micturition occurs when the muscular wall of the bladder contracts and the urethral sphincters relax. It can be assisted by lowering the diaphragm, and contracting the abdominal muscles, thus increasing the pressure within the pelvic cavity. Inhibition of reflex contraction of the bladder is possible for only a limited period of time. Over distension of the bladder is extremely painful and when a painful degree of distension is reached there is a tendency for involuntary relaxation of the sphincters to occur and a small amount of urine to escape, provided there is no mechanical obstruction present.

13. The Skin

The skin is one of the most active organs of the body. It contains the nerve endings of many of the sensory nerves. It is one of the main excretory organs. It plays an important part in the regulation of the body temperature, and it protects the deeper organs from injury and the invasion of micro-organisms.

Structure of the Skin

The skin is composed of two main parts:
- the epidermis
- the dermis or corium (true skin).

THE EPIDERMIS

The epidermis is the most superficial part of the skin and is composed of *stratified squamous epithelium* which varies in thickness in different parts of the body. It is thickest on the palms of the hands and soles of the feet. There are no blood vessels in the epidermis, but its deeper layers are bathed in interstitial fluid which is drained away as lymph.

There are several layers of cells forming the epidermis. They vary in microscopic appearance and are described in two zones:
- the horny zone
- the germinative zone.

The horny zone

This lies superficially and consists of three layers of cells.

The Stratum Corneum or Horny Layer. This is the most superficial layer of cells. The cells have no nuclei and are flat and thin near the surface of the body. The protoplasm in the cells has been changed into a horny substance known as *keratin* which protects the deeper layers of cells. These cells are continually being cast off by friction and being replaced by cells from the deeper layers.

The Stratum Lucidum. This is composed of cells with clear protoplasm and in some of the cells small flattened nuclei are found. These cells replace the stratum corneum as it is cast off.

The Stratum Granulosum. This is composed of several layers of cells which contain many granules in their protoplasm. These granules are thought to be the first stage in the development of keratin.

The cells move towards the surface to replace the cells of the stratum lucidum. Distinct nuclei can be seen in the cells.

The germinative zone

This forms the deeper layer of epidermis and consists of two layers of cells.

The Prickle Cell Layer. This layer consists of cells of different shapes with a distinct nucleus and a short thorn-like protoplasmic process by which they are joined together.

The Basal Cell Layer. This consists of a layer of columnar-shaped cells with oblong nuclei. These cells reproduce rapidly and are continually replacing those in the more superficial layers. Within the basal layer there are cells known as *melanoblasts* which contain fine granules of a pigment known as *melanin.* It is this pigment which gives the colour to the skin in addition to that caused by the blood supply to the dermis.

Figure 246

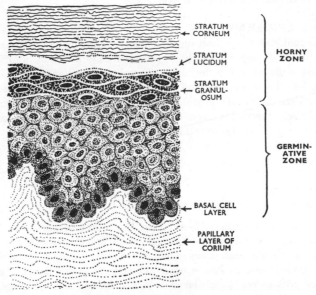

Diagrammatic illustration of the epidermis.

The amount of melanin present varies in individuals and in races. In very dark-skinned people the melanoblasts may be present as superficially as the stratum granulosum.

The maintenance of healthy epidermis is dependent upon three processes being synchronised.

1. Desquamation of the keratinised cells from the surface.

2. Effective keratinisation of cells.

3. Continual cell division in the deepest layers with the pushing of cells towards the surface.

Passing through the epidermis are the hair roots, and the ducts of the sweat glands. The surface of the epidermis is ridged due to the presence of raised projections in the dermis. These projections are the *papillae of the skin*. The ridges formed by the papillae are different in every individual and can be seen clearly at the tips of the fingers and toes. They form the well-known 'finger prints'.

THE DERMIS OR TRUE SKIN

The dermis is tough and elastic. It is composed of *white fibrous tissue* interlaced with *yellow elastic fibres*. In the deeper layers forming the subcutaneous tissue there is areolar and adipose tissue. In the most superficial layer of the dermis there are projections termed the *papillae*.

There are many structures distributed throughout the dermis.

Blood vessels.

Lymphatic capillaries and vessels.

Sensory nerve endings.

Sweat glands and their ducts.

Hair follicles and hairs

Sebaceous glands.

The arrectores pilorum (involuntary muscles attached to the hair follicles).

1. *Blood Vessels*. Within the dermis the arterioles form a fine network with capillary branches supplying the sweat glands, sebaceous glands, hair roots and minute branches passing upwards to the papillae.

2. *Lymphatic Capillaries and Vessels*. There is a network of fine lymphatic capillaries and vessels throughout the dermis which drain lymph from the basal layer of the epidermis and from the dermis.

3. *Sensory Nerve Endings*. Nerve endings which are sensitive to *changes* in *temperature* and *pressure* are widely distributed in the dermis. Excessive changes in either may cause pain. There are no nerve endings in the epidermis.

The skin is an important sensory organ. It is one of the organs through which the individual is aware of his environment.

The sensory nerve endings stimulated in this way become the sensory or cutaneous nerves which convey the impulses through the spinal nerves to the spinal cord and on to the sensory area in the cerebral cortex of the brain, where temperature, touch and pain are perceived.

4. *Sweat Glands*. These are found widely distributed throughout the skin and are most profuse in the palms of the hands, soles of the feet, axillae and groins.

The glands are composed of *epithelial cells* of various shapes, and the *body* of the gland has a coiled appearance. This coil straightens out to form the *duct of the gland* which traverses both the dermis and epidermis to open on to the surface of the skin at a minute depression known as the *pore*. Each gland is supplied by a network of capillaries.

To a minor extent the secretion of sweat provides a route for the excretion of waste materials, but its most important function is in relation to the maintenance of normal body temperature (see p. 305).

Composition of sweat:
water, 99·4 per cent
potassium ⎫
sodium   ⎪
chloride ⎬   0·2 per cent
sulphate ⎭
waste substances, 0·4 per cent.

5. *Hair Follicles*. These consist of a downward growth of epidermal cells into the dermis or even the subcutaneous tissue. At the base of the follicle there is a cluster of cells, called the bulb, from which the hair grows. The hair is formed by the multiplication of cells of the bulb and, as they are pushed upwards and away from their source of nutrition, the cells die and are converted into keratin.

The part of the hair which is below the skin surface is called the root and the part which protrudes from the follicle, the shaft.

The colour of the hair depends on the amount of melanin present. White hair is the result of the replacement of melanin by tiny air bubbles.

6. *The Sebaceous Glands*. These are distributed widely throughout the skin. They are particularly plentiful in the skin of the face and scalp but are absent in the palms of the hands and soles of the feet. The glands are composed of *secretory epithelial cells*, derived from the same epithelium as the hair follicles and arranged to form alveoli. The ducts of the glands open into the hair follicles.

The sebaceous glands secrete an oily substance called *sebum* into the hair follicles. It keeps the hair soft and pliable and gives it a shiny appearance. On the skin it provides some waterproofing, acts as a bactericidal agent preventing the successful invasion of micro-organisms and prevents drying especially on exposure to heat and sunshine.

7. *The Arrectores Pilorum*. These are minute bundles of involuntary muscle fibres which are connected with the hair follicles. When these

Figure 247

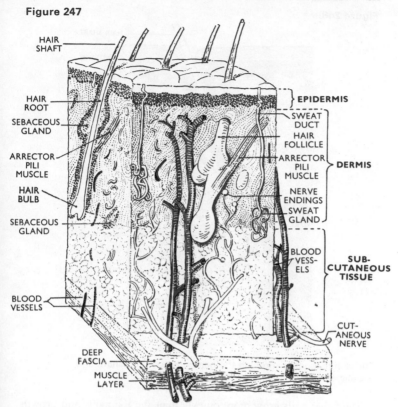

HAIR
SHAFT

HAIR
ROOT

SEBACEOUS
GLAND

ARRECTOR
PILI
MUSCLE

HAIR
BULB

SEBACEOUS
GLAND

BLOOD
VESSELS

DEEP
FASCIA

MUSCLE
LAYER

EPIDERMIS

SWEAT
DUCT

HAIR
FOLLICLE

ARRECTOR
PILI
MUSCLE

NERVE
ENDINGS

SWEAT
GLAND

BLOOD
VESS-
ELS

CUT-
ANEOUS
NERVE

DERMIS

SUB-
CUTANEOUS
TISSUE

The true skin showing blood vessels, hairs, and glands.

muscles contract they make the hair stand erect. This also causes the skin around the hair to become elevated giving the appearance of 'goose flesh'. The muscles are stimulated by sympathetic nerve fibres in fear and in response to cold. Although each muscle is very small the contraction of a large number generates an appreciable amount of heat.

THE NAILS
The nails in human beings are equivalent to the claws, horns and hoofs of animals. They are derived from the same cells as epidermis and hair and consist of a hard, horny type of keratinised dead cells. They protect the tips of the fingers and toes.

The root of the nail is embedded in the skin and is covered by the *cuticle.*

The body of the nail is the exposed part and grows out from the *nail bed* which is the germinative zone of the epidermis. The nail grows from the root which forms the 'half-moon' or *lunula*. The distal free edge of the nail is known as the *free border*.

Figure 248

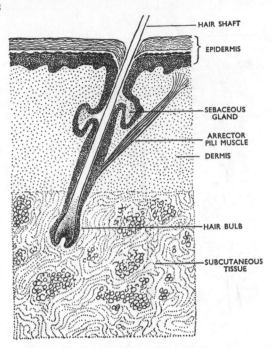

HAIR SHAFT

EPIDERMIS

SEBACEOUS GLAND

ARRECTOR PILI MUSCLE

DERMIS

HAIR BULB

SUBCUTANEOUS TISSUE

The skin magnified showing the hair bulb, root and shaft, and associated sebaceous glands and arrector pili muscle.

The finger nails grow more quickly than the toe nails, and growth is much quicker in the summer than during the winter.

Figure 249

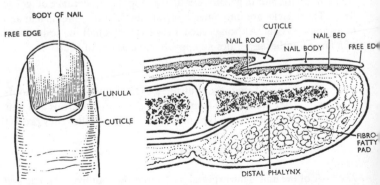

BODY OF NAIL

FREE EDGE

LUNULA

CUTICLE

CUTICLE

NAIL ROOT

NAIL BODY

NAIL BED

FREE EDGE

FIBRO-FATTY PAD

DISTAL PHALYNX

A nail.

(A) showing the body and lunula.

(B) showing a magnified side view of the nail

Additional Functions of the Skin

Protection

The skin is one of the main protective organs of the body. It protects the deeper and more delicate organs, and acts as the main barrier against the invasion of micro-organisms and other harmful agents. Any break in the skin should therefore be cleansed and covered immediately.

Due to the presence of the sensory nerve endings the body reacts by reflex action to unpleasant or painful stimuli, and thus is protected from further injury.

Formation of vitamin D_3

Present in sebum is a fatty substance known as *7-dehydrocholesterol*. The ultra violet rays of the sun have a direct action on 7-dehydrocholesterol producing vitamin D. The vitamin D thus formed is absorbed into the blood stream and is utilised within the body to ensure the satisfactory development and maintenance of bone tissue and the proper utilisation of calcium and phosphorus.

Regulation of Body Temperature

Human beings are warm-blooded animals and the body temperature is maintained at an average of $36C°$ (98·4F). In health, variations are usually limited to between 0·5 and $0·75C°$. It may be found that the temperature in the evening is a little higher than in the morning.

To ensure this constant temperature a fine balance is maintained between heat production in the body and heat lost to the environment.

HEAT PRODUCTION

Heat is produced by several body activities.

1. The Muscles. Muscle contraction produces a large amount of heat. The more strenuous the muscular exercise the greater the heat produced. Shivering involves muscle contraction and produces heat when there is the risk of the body temperature falling below normal.

2. The Liver. It will be remembered that the liver performs many chemical activities each involving the production of heat.

3. The Digestive Organs. Heat is produced by the contraction of the muscle of the alimentary tract and by the chemical reactions involved in digestion.

4. Utilisation of Foodstuffs. The utilisation of carbohydrate and fat within the body produces energy and, as a by-product, heat.

In these and other minor ways the body can produce its own heat, but if heat is continually being produced then a certain amount of heat loss must take place to maintain the normal temperature.

HEAT LOSS

Heat is lost from the body in several ways:

    97 per cent by the skin.
    2 per cent in expired air
    1 per cent in urine and faeces.

Only the heat lost by the skin can be regulated to maintain a constant body temperature. The heat lost by the other routes is obligatory.

Heat loss from the body is affected by the difference between body and environmental temperature, the amount of the body surface exposed to the air and the type of clothes worn. Air is a poor conductor of heat and when layers of air are trapped in the clothing and between the skin and the clothing they act as effective insulators againt excessive heat loss. It has been said that several layers of light weight clothes provide more effective insulation against a low environmental temperature than one heavy garment.

Nervous control

The centre controlling temperature is situated in the *cerebrum* and involves a group of nerve cells in the *hypothalamus*. These nerve cells are described as the *heat regulating centre*. There is also a group of nerve cells in the *medulla oblongata* known as the *vaso motor centre* which controls the calibre of the blood vessels, especially the small arteries and the arterioles.

The heat regulating centre and vaso motor centre are thought to be extremely sensitive to the temperature of the blood and any significant *change* stimulates them to activity. From these centres sympathetic nerves convey impulses to the *sweat glands* and *blood vessels* in the skin.

Activity of the sweat glands

If the temperature of the body is increased by 0·25 to 0·5C° the sweat glands are stimulated to secrete sweat. The sweat secreted is conveyed to the surface of the body by ducts, and this moisture immediately *evaporates* into the atmospheric air, thus *cooling the body*. It cools the body because the heat which evaporates the water is taken from the skin. If the atmospheric air is humid then evaporation of sweat does not take place so readily and beads of sweat appear on the surface of the body. If the atmospheric temperature is high and exercise is being taken the sweat may drip from the body.

Loss of heat from the body by *evaporation* is described as occurring by:

    insensible water loss
    sweating.

In *insensible water loss* heat is being continuously lost by evaporation, even although the sweat glands are not active. Water diffuses upwards from the deeper layers of the skin to the surface of the body and evaporates into the atmospheric air.

In *sweating* the sweat glands are active and secrete sweat on to the surface of the body which evaporates and, in the process, cools the skin.

Effects of vaso-dilation

The amount of heat lost from the skin depends to a great extent on the amount of blood in the vessels which lie in the dermis. As the amount of heat produced in the body increases the arterioles become dilated and more blood pours into the capillary network in the skin. In addition to increasing the amount of sweat produced the *temperature of the skin is raised*. When this happens there is an increase in the amount of heat lost by:

radiation
conduction
convection.

In *radiation* the exposed parts of the body radiate heat away from the body.

In *conduction* the clothes in contact with the skin conduct heat away from the body.

In *convection* the air passing over the exposed parts of the body is heated and rises, cool air replaces it and convection currents are set up. Heat is also lost from the clothes by convection.

If the external environmental temperature is low, for example in winter, or if heat production is decreased, the blood vessels, under the influence of the sympathetic nerves, constrict thus decreasing the blood supply to the skin and so preventing heat loss.

In man, therefore, this fine balance of heat production and heat loss must continuously be maintained to ensure no drastic change in body temperature.

14. The Nervous System

In a systematic study of anatomy and physiology the systems of the body are described separately, but it must be appreciated that they are entirely dependent on each other and no integration to achieve co-ordinated function is possible without the association of each system with the others. The dependence upon each other of the digestive, circulatory and respiratory systems has already been demonstrated.

The two systems primarily involved in the co-ordination of bodily function are the nervous and endocrine systems. The endocrine system will be dealt with later.

The nervous system is composed of large numbers of nerve cells, their processes and a special type of connective tissue called *neuroglia*. The unit of the nervous system is known as a *neurone*.

A neurone consists of:
the nerve cell
the processes of the nerve cell; an *axon* and *dendrites*

THE NERVE CELLS
The nerve cells vary considerably in shape and size. In fact some of the nerve cells are the largest cells in the body. The protoplasm of nerve cells contain many granules and is very highly specialised. The nerve cells form the *grey matter* of the nervous system and are found at the *periphery* of the brain, in the *centre* of the spinal cord and in groups called ganglia outside the brain and spinal cord and as single cells in the walls of organs. Every never cell has one or more processes.

AXONS AND DENDRITES
The axons and dendrites are the processes of the nerve cells and form the *white matter* of the nervous system. They are found *deep* within the brain, at the *periphery* of the spinal cord and as the *peripheral nerves*.

Axons

The axons are the processes or nerve fibres which carry impulses *away* from nerve cells. They are usually longer than the dendrites, and may be as long as 100 cm (approximately 40 inches).

The Structure of an Axon

1. The central portion consists of extremely fine fibres known as the *axis-cylinder*.

2. Surrounding the axis-cylinder is a fatty sheath, the *medullary* or myelin sheath. This gives the nerve a white appearance. Not all nerve fibres have a medullary sheath but when it is present the fibre is described as *medullated* or *myelinated*.

The functions of the myelin sheath are:

(*a*) to act as an insulator

(*b*) to protect the axis-cylinder from pressure or injury.

(*c*) to speed up the flow of nerve impulses through the axon

(*d*) maybe to supply nutrients to the axis cylinder.

Figure 250

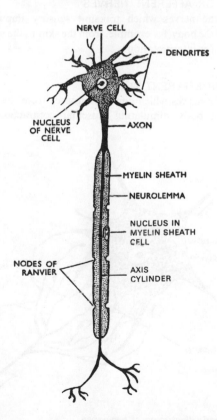

NERVE CELL

DENDRITES

NUCLEUS OF NERVE CELL

AXON

MYELIN SHEATH

NEUROLEMMA

NUCLEUS IN MYELIN SHEATH CELL

NODES OF RANVIER

AXIS CYLINDER

A neurone.

The myelin sheath is absent at intervals of about 1 mm. These spaces are called the *Nodes of Ranvier*. This arrangement is necessary for the transmission of nerve impulses along medullated fibres.

3. *The neurolemma* is a very fine, delicate membrane which surrounds the myelin sheath. At frequent intervals, between the myelin sheath and the neurolemma, nuclei can be seen which are surrounded by protoplasm. The neurolemma is not present in the medullated fibres in the spinal cord and brain. It is only found surrounding the myelin sheath in the peripheral nerves.

Dendrites

The dendrites are the processes or nerve fibres which carry impulses *towards* nerve cells. Generally speaking they are much shorter than the axons but have the same structure.

Nerves are generally divided into two types, *sensory* and *motor*.

SENSORY OR AFFERENT NERVES

These are the nerves which transmit sensory impulses from the periphery of the body, for example, from the skin to the spinal cord then to the brain for interpretation.

MOTOR OR EFFERENT NERVES

These are the nerves which convey impulses from the brain to other parts of the body stimulating muscular contraction or glandular secretion.

Figure 251

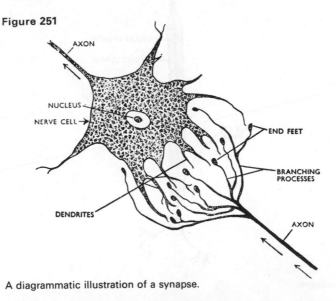

A diagrammatic illustration of a synapse.

MIXED NERVES

In the spinal cord there are separate and distinct pathways for the sensory and motor nerves, but outside the spinal cord they are together enclosed in specialised connective tissue and are known as *mixed nerves*. Large nerves, such as the sciatic, radial and vagus nerves are described as mixed nerves, due to the presence of both sensory and motor fibres in the same sheath.

THE SYNAPSE AND CHEMICAL TRANSMITTERS

There is always more than one neurone involved in the transmission of a nerve impulse from its origin to its effector organ, whether it is a sensory or motor pathway. There is no *anatomical continuity* between these neurones. At its free end the axon of one neurone breaks up into minute branches which terminate in small swellings called *end-feet* which are in close proximity to the dendrites or the cell body of the next neurone. When the nerve impulse reaches the end-feet a chemical substance is released which stimulates the next neurone. There are a number of different substances known to function in this way which are called *chemical transmitters*. Their action is very short lived as immediately they have stimulated the next neurone they are neutralised by an enzyme and rendered inactive. A knowledge of the action of different chemical transmitters has become more important because of the drugs which are available to neutralise them.

The chemical transmitters and their modes of action in the brain and spinal cord are not yet fully understood but it is believed they include *noradrenalin, 5-hydroxytryptamine (serotonin)*, and *gamma amino-butyric acid (GABA)*.

Figure 252

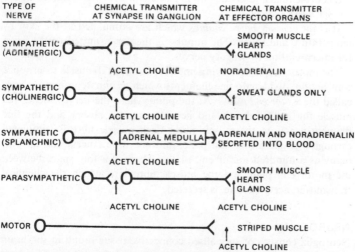

TYPE OF NERVE	CHEMICAL TRANSMITTER AT SYNAPSE IN GANGLION	CHEMICAL TRANSMITTER AT EFFECTOR ORGANS

SYMPATHETIC (ADRENERGIC) — ACETYL CHOLINE — SMOOTH MUSCLE / HEART / GLANDS — NORADRENALIN

SYMPATHETIC (CHOLINERGIC) — ACETYL CHOLINE — SWEAT GLANDS ONLY — ACETYL CHOLINE

SYMPATHETIC (SPLANCHNIC) — ACETYL CHOLINE — ADRENAL MEDULLA — ADRENALIN AND NORADRENALIN SECRETED INTO BLOOD

PARASYMPATHETIC — ACETYL CHOLINE — SMOOTH MUSCLE / HEART / GLANDS — ACETYL CHOLINE

MOTOR — STRIPED MUSCLE — ACETYL CHOLINE

Diagrammatic representation of some nervous system chemical transmitters.

Figure 252 summarises the chemical transmitters which are known to function outside the brain and spinal cord. This diagram should also be used later in conjunction with further study of the nervous system.

Figure 253

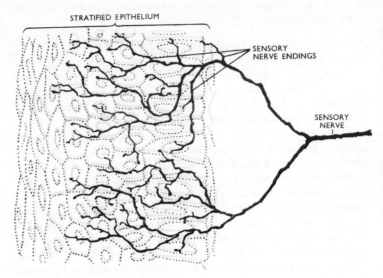

Sensory nerve endings in the skin.

TERMINATION OF NERVES

The sensory nerves, for example, in the skin, lose their myelin sheath and neurolemma and divide into fine branching filaments known as the *sensory nerve endings.*

It is the *sensory nerve endings* which are stimulated in the skin by temperature and touch. The impulse is then transmitted to the brain for interpretation by sensory nerves.

The motor nerves conveying impulses to skeletal muscle to produce contraction, divide into fine filaments which terminate in minute pads called the *motor end plates.* At the point where the nerve reaches the muscle the myelin sheath and neurolemma are absent and the fine filament passes to the muscle fibre. Each muscle fibre is stimulated through a single motor end plate, and one motor nerve can have as many as a hundred motor end plates. There is a tiny space between the motor end plate and the muscle fibre into which the chemical transmitter, acetylcholine, is secreted.

NEUROGLIA

Neuroglia consists of specialised connective tissue found in the brain and spinal cord supporting the nerve cells and their fibres. There are

Figure 254

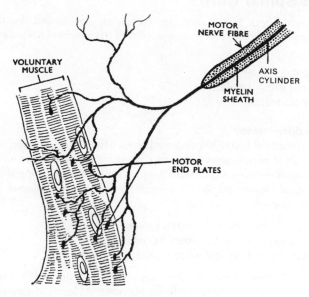

Termination of a motor nerve in muscle showing motor end plates.

three types of neuroglial cells called oligodendrocytes, astrocytes and microglia. Some of these cells are part of the reticulo-endothelial system.

THE PROPERTIES OF NERVE TISSUE

Nerve tissue has the characteristics of *irritability* and *conductivity*.

Irritability is the power to respond to stimulation. In the body this stimulation may be described as partly electrical and partly chemical.

Conductivity means the ability to transmit an impulse. The impulse may be transmitted from:

1. one part of the brain to another
2. the brain to striated muscle resulting in muscle contraction
3. muscles and joints to the brain, contributing to the maintenance of balance
4. the brain to organs of the body resulting in the contraction of smooth muscle or the secretion of glands
5. organs of the body to the brain in association with the regulation of body functions
6. the outside world to the brain through sensory nerve endings in the skin which are stimulated by temperature and touch
7. the outside world to the brain through the special sense organs, i.e., eyes, ears, nose, tongue.

The Membranes Covering the Brain and Spinal Cord

The brain and spinal cord are completely surrounded by three membranes known as the *meninges*. Named from without inwards they are:

the dura mater
the arachnoid mater
the pia mater.

The dura mater

This consists of two layers of dense fibrous tissue. The outer layer takes the place of periosteum on the inner surface of the skull bones and the inner layer provides a protective covering for the brain and spinal cord.

The two layers are closely adherent except where the inner layer sweeps inwards between the:

cerebral hemispheres—the falx cerebri
cerebellar hemispheres—the falx cerebelli
cerebrum and the cerebellum—the tentorium cerebelli.

It will be remembered that the venous blood from the brain is drained into venous sinuses. These are formed between the layers of dura mater where they are separated. The *superior saggital sinus* is formed by the falx cerebri and the tentorium cerebilli forms the *straight* and *transverse sinuses*.

The dura mater continues downwards to line the vertebral canal although it does not replace the periosteum of the vertebrae. It continues beyond the end of the spinal cord (L1, 2) and finally merges with the periosteum of the coccyx.

The arachnoid mater

This is a delicate serous membrane situated between the dura mater and the pia mater. It is separated from the dura mater by a potential space known as the *subdural space,* and from the pia mater by a definite space, the *sub-arachnoid space*. Within the sub-arachnoid space flows the cerebro-spinal fluid. The arachnoid mater passes over the convolutions of the brain and accompanies the inner layer of dura mater in the formation of the falx cerebri, tentorium cerebelli and falx cerebelli. It continues downwards to envelop the spinal cord and ends by merging with the dura mater at the level of the second sacral vertebra.

The pia mater

This is a fine vascular membrane consisting mainly of minute blood vessels supported by fine connective tissue. It closely invests the brain completely covering the convolutions and dipping down into each fissure. It continues downwards to invest the spinal cord. Beyond the

Figure 255

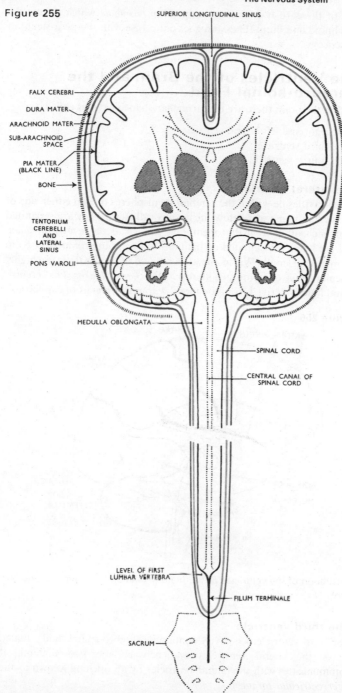

SUPERIOR LONGITUDINAL SINUS

FALX CEREBRI

DURA MATER

ARACHNOID MATER

SUB-ARACHNOID SPACE

PIA MATER (BLACK LINE)

BONE

TENTORIUM CEREBELLI AND LATERAL SINUS

PONS VAROLII

MEDULLA OBLONGATA

SPINAL CORD

CENTRAL CANAL OF SPINAL CORD

LEVEL OF FIRST LUMBAR VERTEBRA

FILUM TERMINALE

SACRUM

The meninges covering the brain and spinal cord.

end of the cord it continues as the *filum terminale* which pierces the arachnoid and dural tubes and goes on to fuse with the periosteum of the coccyx.

The Ventricles of the Brain and the Cerebro-spinal Fluid

Within the brain there are four irregular-shaped cavities called:

the right and left lateral ventricles
the third ventricle
the fourth ventricle

The lateral ventricles

These cavities lie within the cerebral hemispheres one on either side of the median plane just below the corpus callosum. They are separated from each other by a thin membrane known as the *septum lucidum*, and are lined with ciliated epithelium. They are approximately 6 cm (2½ inches) in length. Within the walls of these ventricles there is a highly vascular fringe of membrane known as the *choroid plexus*. The cerebro-spinal fluid is derived from the blood within this plexus of capillaries.

Figure 256

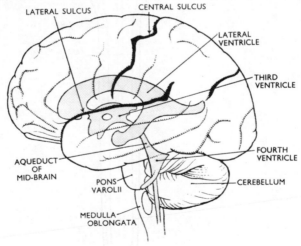

Illustration of the ventricles of the brain in relation to its surface, lateral view.

The third ventricle

The third ventricle is a cavity containing cerebro-spinal fluid situated below the lateral ventricles and between the two thalami. It communicates with the lateral ventricles by an opening known as the *interventricular foramen*.

The fourth ventricle

The fourth ventricle is a lozenge-shaped cavity containing cerebro-spinal fluid situated below and behind the third ventricle, in front of the cerebellum and behind the pons varolii. It communicates with the third ventricle above by a canal known as the *aqueduct of the midbrain* and is continuous below with the *central canal* of the spinal cord.

The cerebro-spinal fluid is formed and secreted within the ventricles of the brain by the *choroid plexus*. It is a clear slightly alkaline fluid with a specific gravity of 1007 and consists of water, amino acids, mineral salts and glucose. After filling the lateral ventricles it flows

Figure 257

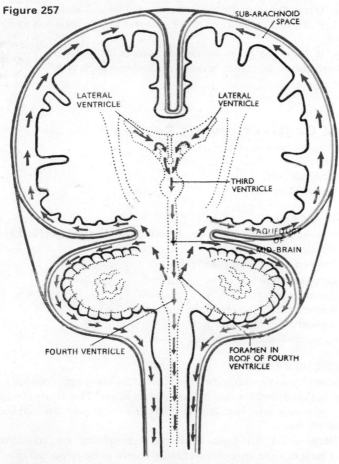

Scheme showing the flow of the cerebral spinal fluid, arrows indicate direction of flow.

through the *interventricular foramen* into the *third ventricle* and from there through the *aqueduct of the midbrain* into the *fourth ventricle*. From the roof of the fourth ventricle the cerebro-spinal fluid flows through two foramina, the *median and lateral foramina* into the *sub-arachnoid space* and subsequently completely surrounds the brain and spinal cord. It also flows from the floor of the fourth ventricle downwards through the central canal of the spinal cord.

The cerebro-spinal fluid is reabsorbed into blood capillaries in the arachnoid mater, and in this way is returned to the circulating blood.

FUNCTIONS OF THE CEREBRO-SPINAL FLUID

1. It supports and protects the delicate structures of the brain and spinal cord.

2. It maintains a uniform pressure around these delicate structures.

3. It acts as a cushion and shock absorber for the brain and spinal cord.

4. It keeps the brain and spinal cord moist and there may be interchange of substances between the fluid and nerve cells.

The Central Nervous System

The central nervous system consists of:

the brain
the spinal cord
the peripheral nerves.

THE BRAIN

The brain constitutes about one-fiftieth of the body weight, and lies within the cranial cavity.

The structures forming the brain are:

the cerebrum or fore brain
the midbrain
the pons varolii   } the brain stem
the medulla oblongata
the cerebellum or hind brain.

THE CEREBRUM

The cerebrum constitutes the largest part of the brain and is divided by a deep cleft termed the *longitudinal cerebral fissure*. This fissure divides the cerebrum into two distinct parts, *the right and left cerebral hemispheres.*

Deep within the brain these two hemispheres are connected by a mass of white matter (nerve fibres), known as the *corpus callosum.*

The superficial or peripheral part of the cerebrum is composed of nerve cells or grey matter forming *the cerebral cortex.*

The cerebral cortex shows many infoldings or furrows which vary in depth. The exposed areas of the folds are termed *gyri or convolutions.* The furrows between the gyri are known as *sulci or fissures.* By these infoldings the surface area of the cerebrum is greatly increased.

Figure 258

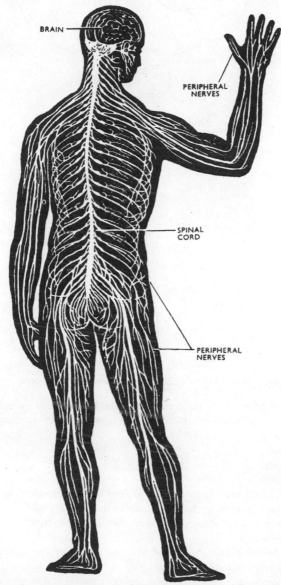

General view of the brain, spinal cord and spinal nerves.

Figure 259

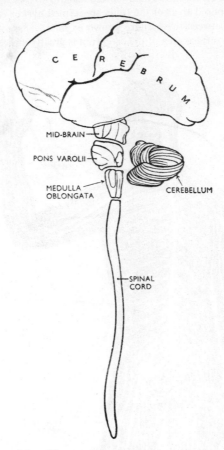

General view of the different parts of the brain and spinal cord.

Each hemisphere of the cerebrum is divided into *lobes* which have been given the names of the bones of the cranium under which they lie:

frontal
parietal
temporal
occipital.

In each hemisphere there are three deep fissures or sulci which play a large part in forming the boundaries of the lobes.

The central sulcus (fissure of Rolando) separates the frontal lobe from the parietal lobe.

The lateral sulcus (fissure of Sylvius) separates the frontal and parietal lobes from the temporal lobe.

The parieto-occipital sulcus separates the parietal and temporal lobes from the occipital lobe.

Figure 260

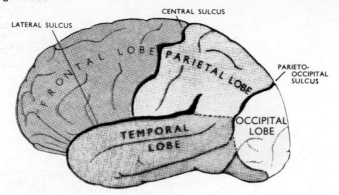

The lobes and sulci of the cerebrum.

Interior of the cerebrum and midbrain

The cerebral cortex is composed mainly of nerve cells. Within the cerebrum the lobes are connected by masses of nerve fibres, or tracts, which make up the white matter of the brain. The fibres which link the different parts of the brain and spinal cord are:

association fibres which connect different parts of the cerebral cortex
   by extending from one gyrus to the next or between adjacent lobes
commisural fibres which connect the two cerebral hemispheres
projection fibres which connect the various parts of the brain with
   one another and continue down through the spinal cord, or are
   nerve fibres passing up from the spinal cord to the cerebral hemispheres.

The *internal capsule* is an important area consisting of projection fibres. All nerve impulses which ascend to and descend from the cerebral cortex are carried by fibres of the internal capsule. These fibres lie deep within the cortex between the basal ganglia and the thalamus.

Functions of the cerebrum

There are three main varieties of activity associated with the functions of the cerebral cortex.

1. The mental activities involved in memory, intelligence, sense of responsibility, thinking, reasoning, moral sense and learning which are attributed to the *higher centres*.

2. Sensory perception which includes the perception of pain, temperature, touch and the *special senses* of sight, hearing, taste and smell.

3. The initiation and control of voluntary muscular contraction.

The precise areas of the brain involved in sensory perception and voluntary muscle contraction are well defined but those associated with mental activities are still the subject of speculation and investigation.

Figure 261

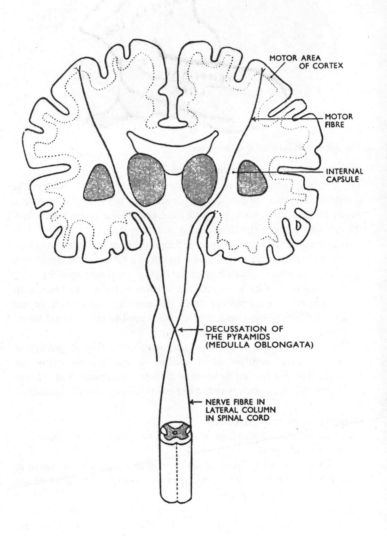

MOTOR AREA
OF CORTEX

MOTOR
FIBRE

INTERNAL
CAPSULE

DECUSSATION OF
THE PYRAMIDS
(MEDULLA OBLONGATA)

NERVE FIBRE IN
LATERAL COLUMN
IN SPINAL CORD

Diagram illustrating an upper motor neurone.

Figure 262

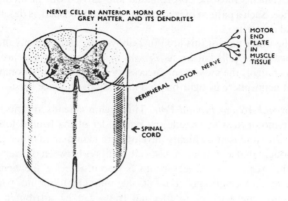

Diagram illustrating a lower motor neurone.

Functional areas of the cerebrum

The Motor or Pre-central Sulcus Area. This region of the cerebral cortex lies in the frontal lobe immediately anterior to the *central sulcus*. The nerve cells in this area are very large and are known as the pyramidal cells. They initiate all voluntary movements. A nerve fibre from a pyramidal cell passes downwards through the internal capsule of the brain as far as the medulla oblongata where it crosses to the opposite side then descends in the spinal cord. At the appropriate level in the spinal cord the nerve impulse crosses a synapse to a second neurone which transmits it to the muscle.

This means that the motor area of the *right hemisphere* controls voluntary muscle movement on the *left side of the body* and vice versa. The neurone with its cell in the cerebrum is called the *upper motor neurone* and the other, with its cell in the spinal cord, the *lower motor neurone.* Damage to either of these neurones may result in paralysis.

In the motor area of the cerebrum the body is represented upside down. The cells in the upper part controlling the feet and the cells in the deepest part controlling the head, neck, face and fingers.

The Pre-motor Area. This region lies in the frontal lobe immediately in front of the motor area. The cells are thought to exert a controlling influence over the motor area, ensuring an orderly series of movements. For example, in tying a shoe lace or writing, many muscles contract but the movements must be co-ordinated and carried out in their proper sequence. Such a pattern of movement, when established, is described as manual dexterity.

In the lower part of this area just above the lateral sulcus there is a group of nerve cells known as *Broca's or the motor speech area.* This region controls the movements necessary for speech. It is dominant in the left hemisphere in right-handed individuals and vice versa.

The Frontal Area or Frontal Pole. This region extends anteriorly from the pre-motor area to include the remainder of the frontal lobe. It is a large area and is more highly developed in man than in other animals. It is thought that a number of association fibres between this region and the other regions in the cerebrum is responsible for the behaviour, character and emotional state of the individual but no particular behaviour, character or intellectual traits can be attributed to the activity of any one group of cells.

Sensory or Post-central Sulcus Area. This is the region of the cerebrum which lies behind the central sulcus. Here sensations of pain, temperature, pressure and touch, all knowledge of muscular movement and the position of joints are received and interpreted. The sensory area of the *right hemisphere* receives and interprets impulses from the *left side of the body* and vice versa.

The Parietal Area. This region lies behind the post-central area and includes the greater part of the parietal lobe. Its functions are thought to be associated with obtaining and retaining accurate knowledge of objects. It has been suggested that objects can be recognised by touch alone because of the knowledge from past experience retained in this area.

Sensory Speech Area. This area is situated in the lower part of the parietal region. It is here that the spoken word is perceived. The sensory speech area extends into part of the temporal lobe. There is a dominant area in the *right hemisphere* if the individual is *left handed* and vice versa.

The Auditory or Hearing Area. This region lies immediately below the lateral sulcus within the temporal lobe. The cells receive and interpret impulses transmitted from the inner ear by the auditory nerve. The auditory area is active in both hemispheres.

The Smell Area. Deep within the temporal lobe are groups of nerve cells associated with the reception and interpretation of impulses received from the olfactory nerve (nerve of the sense of smell).

Figure 263

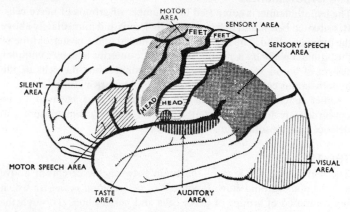

Illustration of the cerebrum showing the functional areas.

The Taste Area. This region is thought to lie just above the lateral sulcus in the deep layers of the sensory area, and it is in this area that the nerve impulses from the tongue are interpreted.

The Visual Area. This region lies behind the parieto-occipital sulcus and includes the greater part of the occipital lobe. The optic nerves or nerves of the sense of sight pass from the eye to this area which receives and interprets the impulses as visual impressions.

Deep within the cerebral hemispheres are groups of nerve cells known as *ganglia or nuclei.* These nuclei act as *relay stations* where synapses occur between neurones. Important masses of grey matter include:
the basal ganglia
the thalamus
the hypothalamus.

The basal ganglia
This area of grey matter, lying deep within the cerebral hemispheres, is thought to influence skeletal muscle tone. If this control is inadequate or absent movements are jerky and clumsy.

The thalamus
The thalamus consists of a mass of nerve cells situated within the cerebral hemispheres just below the corpus callosum.

All sensory nerves from the periphery of the body conveying impulses of pain, temperature, pressure and touch are conveyed by two neurones to the thalamus. It is here that very crude uncritical sensations reach consciousness. The thalamus is unable to distinguish finer sensations. The discrimination and highly critical interpretation of these sensations occur in the sensory area of the cerebral cortex.

The hypothalamus

The hypothalamus is composed of a number of groups of nerve cells. It is situated below and in front of the thalamus and immediately above the pituitary gland. The hypothalamus is linked to the posterior lobe of the pituitary gland by nerve fibres and to the anterior lobe by a complex system of blood vessels. Through these connections the cells of the hypothalamus control the secretion of hormones from both lobes.

Other functions with which the hypothalamus is believed to be concerned are hunger, thirst, the control of body temperature and defence reactions including fear and rage.

THE MIDBRAIN

The midbrain is the area of the brain between the cerebrum above and the pons varolii below. It consists of *two cerebral peduncles* which are composed of groups of nerve cells and nerve fibres. Through the midbrain fibres pass up to the cerebrum from the cerebellum and spinal cord, and descend from the cerebrum to the cerebellum and spinal cord.

The nerve cells act as relay stations for the ascending and descending nerve fibres. Two groups of cells of particular note are the *medial and lateral geniculate bodies* which provide cell stations for the transmission of nerve impulses from the optic nerves and the vestibular portion of the auditory nerves to the cerebellum. These nerve impulses play a major part in the maintenance of balance of the body.

THE PONS VAROLII

The pons varolii is situated in front of the cerebellum below the

Figure 264

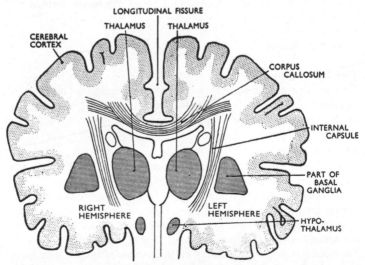

Schematic diagram of the interior of the cerebrum.

midbrain and above the medulla oblongata. It consists mainly of nerve fibres which form a bridge between the two hemispheres of the cerebellum and of fibres passing between the higher levels of the brain and the spinal cord. There are also groups of nerve cells within the pons which act as relay stations, some of which are associated with the cranial nerves.

The anatomical structure of the pons varolii differs from that of the cerebrum in that the *nerve cells lie deeply* and the *nerve fibres lie superficially*.

THE MEDULLA OBLONGATA

The medulla oblongata extends from the pons varolii above and is continuous with the spinal cord below and is about approximately 2·5 cm (1 inch) in length. It is pyramidal in shape being narrow below and broad above. It lies just within the cranium above the foramen magnum. Its anterior and posterior surfaces are marked by a central fissure.

The outer aspect is composed of nerve fibres which are the pathways formed by nerves running to and from the brain and spinal cord.

Grey matter or nerve cells lie *centrally* within the medulla. Some of these cells constitute relay stations for sensory nerves ascending to the cerebrum. Several important cranial nerves arise from nuclei in the medulla.

The so-called *vital centres*, associated with autonomic reflex activity, are contained within its deeper structures, these are:

the cardiac centre
the respiratory centre
the vaso-motor centre
the reflex centres of vomiting, coughing, sneezing and swallowing.

Functions

1. In the medulla oblongata the majority of *motor nerves* descending from the motor area of the cerebral cortex to the spinal cord cross over from left to right and right to left, therefore the cerebral hemispheres control all muscular movement on the opposite side of the body. The area where the fibres cross is known as the *decussation of the pyramids*.

2. Some of the *sensory nerves* ascending to the cerebrum from the spinal cord also cross over from left to right and vice versa, and this forms the *sensory decussation*.

3. *The cardiac centre* controls the rate and force of cardiac contraction. Sympathetic and parasympathetic nerve fibres pass from the medulla to the heart. It will be remembered that the sympathetic stimulation increases the rate and force of the heart beat and para-sympathetic stimulation has the opposite effect.

4. *The respiratory centre* controls the rate and depth of respiration. From this centre nerve impulses stimulate the phrenic and intercostal

nerves which in turn are responsible for stimulating contraction of the diaphragm and intercostal muscles, thus initiating the mechanism of respiration. It will be remembered that the respiratory centre is stimulated by excess carbon dioxide and lack of oxygen in the blood.

Figure 265

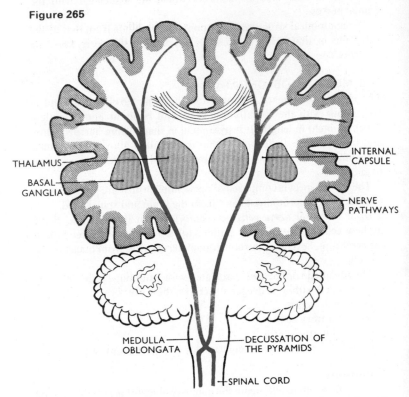

Scheme showing motor fibres passing through the internal capsule and crossing at the decussation of the pyramids.

5. *The vaso-motor centre* controls the calibre of the blood vessels, especially the small arteries and arterioles which have a large proportion of smooth muscle fibres in the tunica media. Cerebral blood vessels are not affected.

Vaso-motor impulses reach the blood vessels through the sympathetic part of the autonomic nervous system. Stimulation causes constriction of all blood vessels except the coronary vessels which are dilated. Vaso-dilation is the result of lack of stimulation.

The sources of stimulation of the vaso-motor centre are the arterial baroreceptors, emotions such as sexual excitement and anger. Pain usually causes vaso-constriction although severe pain may cause vaso-dilation, a fall in blood pressure and fainting.

6. When irritating substances are present in the stomach or respiratory tract nerve impulses pass to the medulla oblongata. These impulses stimulate the *reflex centres* which initiate the reflex actions of vomiting, coughing or sneezing.

THE CEREBELLUM

The cerebellum is situated behind the pons varolii and immediately below the posterior portion of the cerebrum and lies within the posterior cranial fossa. It is ovoid in shape and presents two hemispheres which are separated by a narrow median strip known as the *vermis*. Grey matter is found forming the surface of the cerebellum, and the white matter lies deeply, presenting a branched appearance termed the *arbor vitae*.

The nerve fibres which enter and leave the cerebellum do so by three pathways known as the cerebellar peduncles.

1. *The superior cerebellar peduncles* connect the cerebellum with the midbrain and cerebrum.

2. *The middle cerebellar peduncles* connect the cerebellum with the pons varolii.

3. *The inferior cerebellar peduncles* connect the cerebellum with the medulla oblongata and the spinal cord.

Functions

The cerebellum is concerned with voluntary muscular movement and balance.

Figure 266

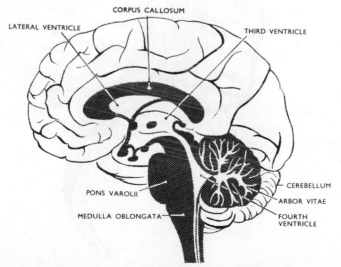

The cerebellum in relation to the pons varolii and cerebrum.

Cerebellar activities are carried out *below the level of consciousness* and are, therefore, not under the control of the will.

1. It is considered to be responsible for the maintenance of muscular tone.

2. It influences the various muscle groups engaged in voluntary muscular movement. This means that it controls and co-ordinates the movements of various groups of muscles ensuring smooth, even, and precise actions.

3. The cerebellum coordinates activities associated with the maintenance of the balance and equilibrium of the body in space. The sensory input is derived from the muscles and joints, the eyes and the ears. Impulses from the muscles and joints indicate their position in relation to the body as a whole and those from the eyes and the semi-circular canals in the ears provide information about the position of the

Figure 267

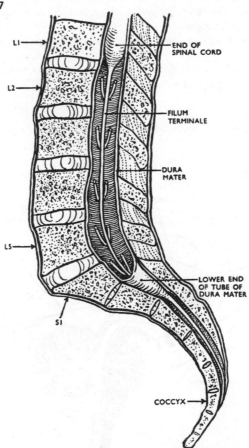

Diagram illustrating the termination of the spinal cord.

head in space. Impulses from the cerebellum influence the contraction of skeletal muscle so that balance of the body is maintained.

Damage to the cerebellum results in clumsy unco-ordinated muscular movement, staggering gait and inability to carry out smooth, steady precise actions. The muscles may become flaccid due to loss of tone.

THE SPINAL CORD

The spinal cord is the elongated almost cylindrical part of the central nervous system. which lies within the neural canal of the vertebral column. It is continuous above with the medulla oblongata and extends from the *upper border of the atlas* to the lower border of the *first lumbar vertebra.*

The spinal cord is approximately 45 cm (18 inches) in length, and is about the thickness of a little finger. It is surrounded by the dura, arachnoid and pia maters as described previously. Cerebro-spinal fluid is present in the central canal of the spinal cord and in the sub-arachnoid space. When a specimen of cerebro-spinal fluid is required it is taken at a point beyond the end of the cord, i.e., below the level of the second lumbar vertebra.

Figure 268

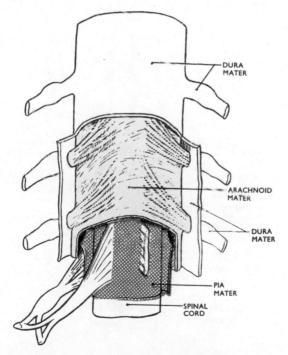

Illustration showing the coverings of the spinal cord.

STRUCTURE

The spinal cord is the nervous tissue link between the brain and the

Figure 269

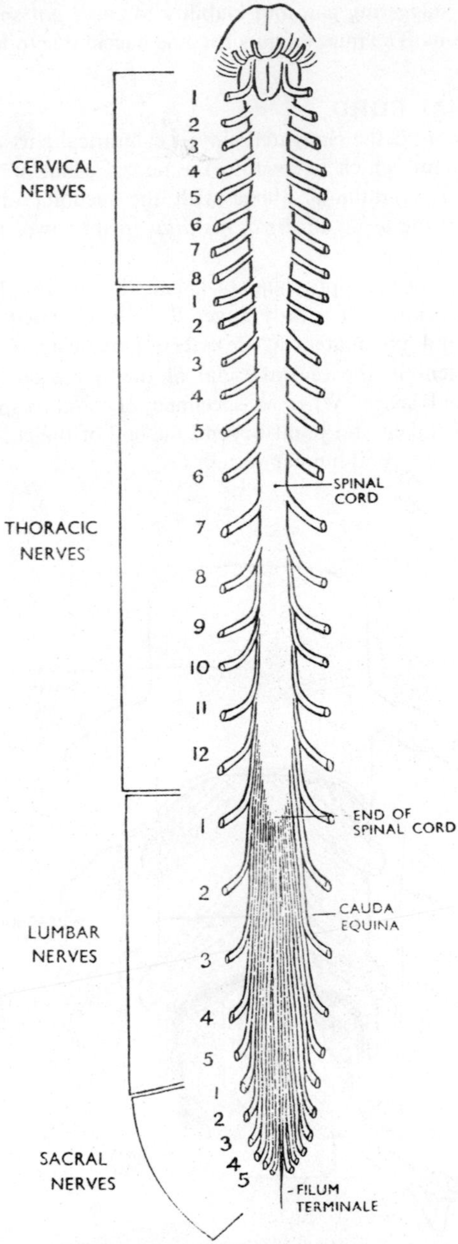

CERVICAL
NERVES

1
2
3
4
5
6
7
8

THORACIC
NERVES

1
2
3
4
5
6
7
8
9
10
11
12

SPINAL
CORD

END OF
SPINAL CORD

CAUDA
EQUINA

LUMBAR
NERVES

1
2
3
4
5

SACRAL
NERVES

1
2
3
4
5

FILUM
TERMINALE

Diagram showing the spinal nerves.

organs of the body. Nerves conveying impulses from the brain to the various organs of the body descend through the spinal cord. At the appropriate level they leave the cord and pass to the organ which they supply. Similarly, sensory nerves from the skin and other organs enter and pass upwards in the spinal cord to the brain.

Figure 270

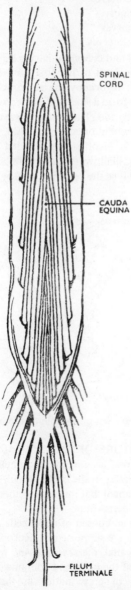

SPINAL CORD

CAUDA EQUINA

FILUM TERMINALE

Scheme showing the formation of the cauda equina.

Thirty-one pairs of nerves arise from the spinal cord which are known as the *spinal nerves*. Each nerve is described as having an *anterior and posterior nerve root*. The pairs of spinal nerves are named according to the part of the neural canal from which they emerge:

8 pairs of cervical nerves

12 pairs of thoracic nerves

5 pairs of lumbar nerves

5 pairs of sacral nerves

1 pair of coccygeal nerves.

The lumbar, sacral and coccygeal nerves leave the *spinal cord* before its termination at the level of the first lumbar vertebra, and extend vertically downwards to occupy the sub-arachnoid space below this level. In this way they form a sheath of nerves which resembles a horse's tail called the *cauda equina*. These nerves leave the *neural canal* at the appropriate lumbar, sacral or coccygeal level.

The spinal cord is incompletely divided into two equal parts, anteriorly by a short, shallow *median fissure* and posteriorly by a deep narrow septum known as the *posterior median septum*.

Figure 271

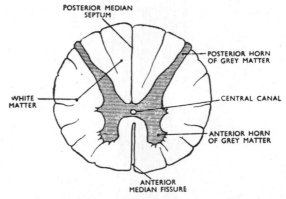

Illustration of a cross section of the spinal cord at the thoracic level.

INTERNAL STRUCTURE OF THE SPINAL CORD

Examination of a cross section of the spinal cord shows that it is composed of grey matter in the centre which is surrounded by white matter. It should be noted that this arrangement is the opposite to that in the cerebrum and cerebellum.

The grey matter is composed of nerve cells and the white matter of nerve fibres and both are supported by neuroglia. Within the centre of the spinal cord is a canal, the *central canal*, which is continuous with the fourth ventricle of the brain and contains cerebro-spinal fluid.

The arrangement of grey matter in the spinal cord bears a resemblance to the letter H. The points of the letter situated in the anterior aspect are termed the *anterior horns of grey matter* and the points lying posteriorly are described as the *posterior horns of grey matter*. The area of grey matter lying transversely is known as the transverse commissure and it is pierced by the central canal.

The grey matter of the spinal cord consists of nerve cells which may:
1. receive sensory impulses from the periphery of the body
2. be the cells of lower motor neurones which transmit impulses to the skeletal muscles
3. be the cells of connector neurones which link the sensory and motor neurones in the formation of spinal reflex arcs.

At each point where nerve impulses are passed from one neurone to another there is a synaptic gap and a chemical transmitter.

The posterior horns of grey matter

These are composed of nerve cells which are stimulated by *sensory impulses* from the periphery of the body. The nerve fibres of these cells, which contribute to the formation of the white matter of the cord, transmit the sensory impulses to the brain.

Figure 272

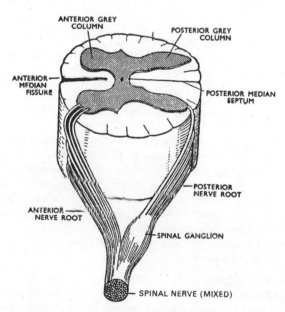

A section of the spinal cord and nerve roots showing the spinal ganglion.

The anterior horns of grey matter

These are composed of the cells of the lower motor neurones which are stimulated by the axons of the upper motor neurones.

White matter

The white matter of the spinal cord is arranged in three definite tracts or columns:

the anterior columns
the posterior columns
the lateral columns.

These columns are formed by sensory nerve fibres *ascending* to the brain from the periphery of the body, motor nerve fibres *descending from* the brain prior to reaching the organs or muscles which they stimulate and the fibres of connector neurones.

The posterior root ganglia or spinal ganglia

These ganglia are composed of nerve cells which lie just outside the spinal cord on the pathway of the sensory nerves. They are the cells of the sensory nerve fibres conveying impulses to the cord from the periphery of the body. The fibres of these cells form the posterior nerve root.

Pathways or nerve tracts in the spinal cord

There are two main tracts to be considered:

sensory tracts
motor tracts.

Sensory nerve tracts (ascending)

General sensations are appreciated by all parts of the body. These sensations may be described as deep and superficial.

Deep sensations include deep pain, pressure and the position of muscles and joints.

Superficial sensations include light touch, pain, temperature and pressure.

All these sensations must reach the sensory area of the cerebral cortex. It will be recalled that sensations from the left side of the body are received in the right sensory area of the cortex and vice versa.

The sensory nerves commence in sensory end organs, for example, in the skin. When the nerve-endings are stimulated impulses are transmitted to the posterior root ganglion and from there the impulses are conveyed into the posterior horns of grey matter in the spinal cord. Depending upon the type of stimuli, the nerve fibres will ascend in one of the following sensory nerve tracts.

1. *Posterior Columns or Tracts.* The sensory nerves which ascend in these columns transmit impulses of light touch from the skin and the sense of position from joints and muscles. In the medulla oblongata a synapse occurs with groups of nerve cells which act as relay stations. The nerve fibres now cross over to the opposite side and this crossing is termed the *sensory decussation*. The impulses are then transmitted to the *thalamus* and, after another synapse, to the sensory area of the cerebral cortex where they are interpreted.

2. *Lateral Columns or Tracts.* The sensory nerves which ascend in these columns convey the impulses of pain, temperature and pressure. The nerves after leaving the posterior root ganglion enter the posterior horn of grey matter in the spinal cord where a synapse takes place. The impulses are them transmitted across the spinal cord from the left side to the right side, and vice versa, and conveyed upwards through the lateral columns of the opposite side to the thalamus and thus to the sensory area of the cerebral cortex where perception occurs.

Figure 273

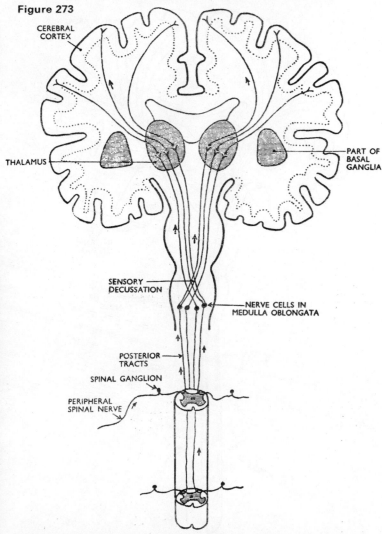

Illustration of some of the sensory pathways from a peripheral nerve to the sensory area of the cerebrum.

The non-sensory impulses of muscle and joint position are also conveyed in the lateral columns. These impulses are transmitted to the cerebellum which carries out its function of muscular co-ordination in the maintenance of balance.

It will be observed that there are three relay stations in the pathways

Figure 274

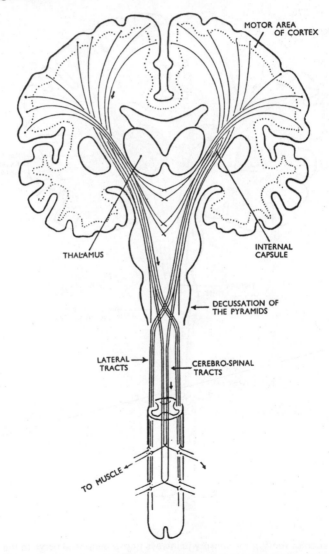

Motor pathway in the cerebrum, medulla oblongata and spinal cord.

of the sensory nerves. The nerves ascending in the *posterior columns of white matter* have relay stations:

in the posterior root ganglion
in the medulla oblongata
in the thalamus.

The nerves ascending in the *lateral columns of white matter* have relay stations:

in the posterior root ganglion
in the posterior horn of grey matter in the spinal cord
in the thalamus.

Motor nerve tracts (descending)

The important tracts are the *pyramidal tracts* which lie in the lateral columns of white matter.

The pyramidal cells in the motor area of the cerebrum initiate impulses which are transmitted by axons. These axons descend through the internal capsule, midbrain and pons to the medulla oblongata. In the medulla most of these nerve fibres cross to the opposite side at the *decussation of the pyramids*. The fibres then descend through the lateral columns of white matter and at the appropriate level enter the anterior horn of grey matter where they form synapses with anterior horn cells. The impulses are then conveyed by motor nerves to the muscle. In the muscle the nerves divide into motor end plates which stimulate the muscle fibres to contract.

The terms upper motor neurone and lower motor neurone are frequently used in association with the motor nerve pathways.

The upper motor neurone consists of a nerve cell in the motor area of the cerebral cortex and the nerve fibre which descends until it reaches the appropriate level in the spinal cord.

The lower motor neurone consists of a nerve cell in the anterior horn of grey matter of the spinal cord, its dendrites and the axon which conveys the impulse to the muscle fibre.

REFLEX ACTION

A reflex action can be described as an automatic motor response to a sensory stimulus without the brain being immediately involved.

Most reflex actions occur without consideration on the part of the individual, and are frequently protective in function. There are a number of familiar reflex actions.

1. The quick closing of the eyelid if the eye is touched or when something threatens to touch it. This is known as the *corneal reflex*.

2. The sudden withdrawal of the hand if the fingers touch something hot.

3. The quick recovery of the balance of the body to prevent falling, after a slip.

4. The quick jerk of the leg if the patella tendon is tapped, this is termed the *knee jerk*.

5. A sudden coughing attack if a crumb is inhaled.

These responses to sensory stimuli are instant although the individual is eventually conscious of their occurrence.

Many other reflex actions occur within the body which never reach consciousness, for example, the movements of the stomach and small intestine, the changes in the calibre of the blood vessels, variation in heart and respiration rates and the secretion of glands.

A reflex action can only take place if there is a complete reflex arc.

Figure 275

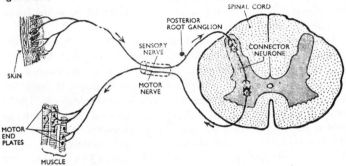

Scheme of a simple reflex arc.

A simple reflex arc has the following three components.

1. *A sensory neurone* which comprises the sensory nerve endings in an organ, the sensory nerve, the posterior root ganglion cell and its nerve fibre which passes to the posterior horn of grey matter in the spinal cord.

2. *A connector neurone* which consists of a nerve cell, its dendrites and axon in the spinal cord.

3. *A motor neurone* consisting of a nerve cell and its dendrites in the anterior horn of grey matter, the axon of this nerve cell and the motor end plates which terminate in muscle.

Physiology of reflex action

The process by which a reflex action is carried out commences when a sensory nerve ending is stimulated. The impulse is conveyed through the sensory nerve to the spinal cord by the sensory neurone which forms a synapse with the dendrites of the connector neurone. The connector neurone may transmit the impulse to the motor neurone at the same level or to a number of motor neurones at different levels, depending on the muscles involved in the motor response. A second synapse occurs where the impulse passes from the connector neurone to the dendrites of the motor neurone. The motor neurone will then convey the impulse to the muscles thus stimulating them to contract.

It must be appreciated that in most reflex actions many reflex arcs are involved, for example, if an individual had his arm or back touched unexpectedly he would jerk the part away, and associated movements of the arm, shoulder and trunk would occur. He would also turn his head to see what had happened and may even utter an exclamation of surprise. These movements would involve a large number of muscles, thus many motor nerves were stimulated although the reflex action was initiated by only a few sensory nerves.

Reflex actions are carried out very rapidly, and even although the brain is not directly involved impulses are transmitted to the cerebrum so that the individual is *aware of what is happening*.

In the infant the act of micturition and defaecation is carried out reflexly. When there is sufficient pressure in the bladder or rectum the sensory nerve endings in their walls are stimulated and the impulse is conveyed through a reflex arc with resultant contraction of the muscles of the bladder and rectum thus expelling the contents.

Inhibition of reflex actions

As the child grows older the sensory nerve tracts of the spinal cord become fully developed, the impulses from the bladder and rectum reach the brain and can be controlled. Thus the act of micturition and defaecation becomes a conscious voluntary act.

Many reflex actions are inhibited thus preventing overactivity of the muscles which otherwise would contract in response to every sensory stimulus, even although a response was unnecessary.

Certain reflex actions cannot be inhibited by the cerebrum, for example, the secretion of glands, vaso-motor, respiratory, cardiac and alimentary reflexes.

If there is damage to the spinal cord in an adult and the sensory nerve pathways are destroyed the stimulus will not reach the sensory area of the brain, therefore the individual will not be conscious of a full bladder or rectum. When the pressure reaches a certain level the bladder or rectum will be emptied by reflex action. This involuntary action is termed *incontinence* of urine and faeces.

THE PERIPHERAL NERVES

The peripheral nerves are composed of sensory nerve fibres conveying impulses from sensory end organs such as the skin, eye and ear to the brain, and motor nerves conveying impulses from the brain through the spinal cord to the effector organs, for example, skeletal muscles. The peripheral nerves are frequently mixed nerves.

The peripheral nerves consist of numerous nerve fibres collected into bundles. Each bundle has several coverings of protective connective tissue.

1. *The endoneurium.* This is a delicate connective tissue which surrounds each individual fibre.

2. *The Perineurium.* This is a smooth connective tissue which surrounds each *bundle* of fibres.

3. *The Epineurium.* If there are many bundles they are completely surrounded by connective tissue known as the epineurium. Most of the large nerves possess this outer protective covering.

The peripheral nerves include:
31 pairs of spinal nerves which arise from the spinal cord
12 pairs of cranial nerves which arise from the brain.

Figure 276

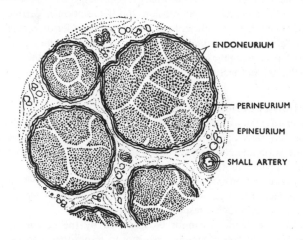

Transverse section of a peripheral nerve showing the various coverings.

THE SPINAL NERVES

There are *thirty-one pairs of spinal nerves.* They are numbered according to the level of the spinal column from which they emerge. They are named and grouped according to the vertebrae with which they are associated.

    8 cervical nerves
   12 thoracic nerves
    5 lumbar nerves
    5 sacral nerves
    1 coccygeal nerve.

It will be recalled that there are 7 cervical vertebrae but 8 cervical nerves. The first cervical nerve arises *above the atlas* and the last one *below the seventh cervical vertebra.*

The spinal nerves arise from both sides of the spinal cord and emerge through the intervertebral foramina. Each nerve is formed by the union of *motor and sensory nerve roots* and is therefore a mixed nerve. Each spinal nerve has a contribution from the sympathetic part of the autonomic nervous system in the form of a *preganglionic fibre.*

Nerve roots

The anterior nerve root consists of *motor nerve fibres* which are the axons of the nerve cells in the anterior horn of grey matter in the spinal cord and, in the thoracic and lumbar regions, sympathetic nerve fibres which are the axons of cells in the lateral horn of greay matter.

The posterior nerve root consists of *sensory nerve fibres* which are characterised by the presence of the *spinal ganglion* or *posterior root ganglion*. The fibres enter the cord at the posterior horn.

For a very short distance after leaving the spinal cord the nerve roots have a covering of dura and arachnoid maters. These coverings terminate before the two roots join to form the spinal nerve. The nerve roots have no covering of pia mater.

Figure 277

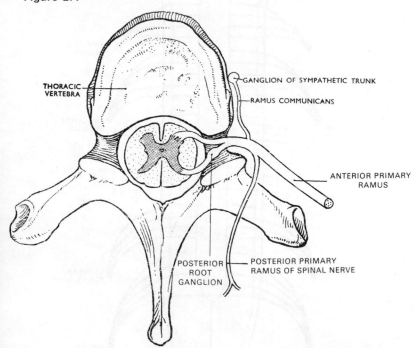

THORACIC VERTEBRA

GANGLION OF SYMPATHETIC TRUNK

RAMUS COMMUNICANS

ANTERIOR PRIMARY RAMUS

POSTERIOR ROOT GANGLION

POSTERIOR PRIMARY RAMUS OF SPINAL NERVE

Diagram illustrating the association between spinal and sympathetic nerves.

Immediately after emerging from the intervertebral foramen each spinal nerve divides into a *posterior and anterior primary ramus* and a *ramus communicantes*. The ramus communicantes is part of the sympathetic portion of the autonomic nervous system.

The Posterior Primary Rami. The posterior primary rami pass backwards and divide into medial and lateral branches to supply the muscles and skin at the posterior aspect of the whole trunk.

The Anterior Primary Rami. The anterior primary rami supply the anterior and lateral aspect of the whole trunk and the upper and lower limbs.

In the cervical, lumbar and sacral regions the anterior primary rami unite near their origins to form a large mass of nerves known as

Figure 278

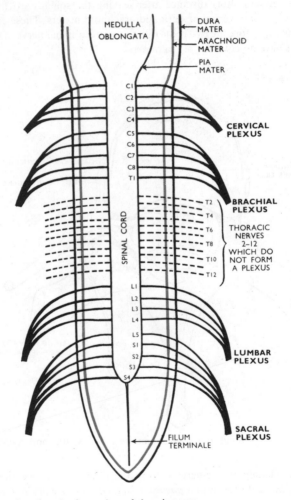

Scheme showing the formation of the plexuses.

a *plexus.* In the plexuses many nerves are regrouped and rearranged before proceeding to supply the skin, muscle and joints of a particular area.

There are five large plexuses formed on each side of the vertebral column. In this discussion the plexus on only one side will be described:

the cervical plexus
the brachial plexus
the lumbar plexus
the sacral plexus
the coccygeal plexus.

The cervical plexus

The anterior primary rami of the *first four cervical nerves* form the cervical plexus.

The plexus lies opposite the 1st, 2nd, 3rd and 4th cervical vertebrae. It lies under the protection of the sternomastoid muscle.

The branches of the cervical plexus are divided into two groups: *superficial and deep*.

Figure 279

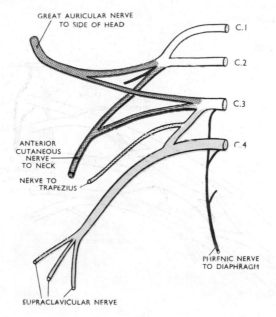

The cervical plexus.

The superficial branches supply the structures at the back and sides of the head and the skin in front of the neck to the level of the sternum.

The deep branches supply muscles of the neck, for example, the sternomastoid and the trapezius.

In addition to the nerve supply to the skin and muscles in the area the *phrenic nerve* arises from the cervical plexus.

The phrenic nerve originates from the C. 3, 4 and 5 and passes downwards through the thoracic cavity in front of the root of the lung to supply the muscle of the diaphragm with impulses which stimulate it to contract.

The brachial plexus

The anterior primary rami of the *lower four cervical nerves* and the *first thoracic nerve* form the brachial plexus.

The plexus is situated above and behind the subclavian vessels and partly in the axilla.

The branches of the brachial plexus supply the skin and muscles of the upper limb and some of the chest muscles. Five large nerves and a number of smaller ones emerge from this plexus, each with a contribution from more than one nerve root:

the circumflex nerve

the radial nerve

the musculocutaneous nerve

the median nerve

the ulnar nerve.

Figure 280

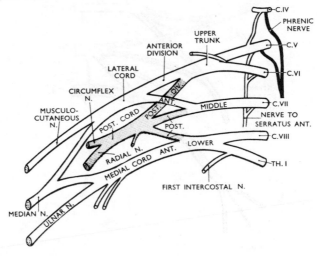

The brachial plexus.

The circumflex nerve winds round the humerus at the level of the surgical neck. It then breaks up into minute branches to supply the deltoid muscle and the shoulder joint.

The radial nerve is the largest branch of the brachial plexus. It winds round the posterior aspect of the humerus in the spiral groove to supply the triceps muscle. It then crosses in front of the elbow joint and passes downwards in the forearm supplying the extensor muscles of the wrist and fingers. It continues into the back of the hand to supply the skin of the thumb, the first two fingers and the lateral half of the ring finger.

The musculocutaneous nerve passes downwards to the lateral aspect of the forearm. It supplies the muscles of the upper arm and the skin of the forearm.

The median nerve passes down the midline of the arm in close association with the brachial artery. It passes in front of the elbow joint then downwards to supply the muscles of the anterior aspect of the forearm. It continues into the hand where it supplies small muscles and the skin of the anterior aspect of the thumb, first two fingers and the lateral half of the third finger. It gives off no branches above the elbow.

Figure 281

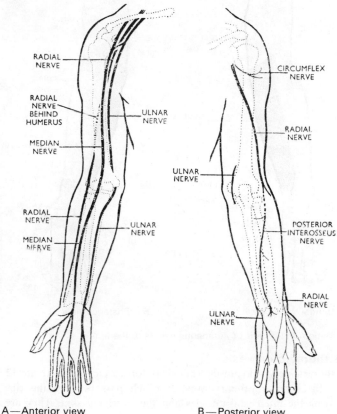

A—Anterior view   B—Posterior view

Main nerves of the arm.

The ulnar nerve descends through the upper arm lying medial to the brachial artery. It passes behind the medial epicondyle of the humerus to supply the muscles on the ulnar aspect of the forearm. It continues to supply the mucles in the palm of the hand and the skin of the whole of the little finger and both sides of the medial half of the ring finger. It gives off no branches above the elbow.

Figure 282

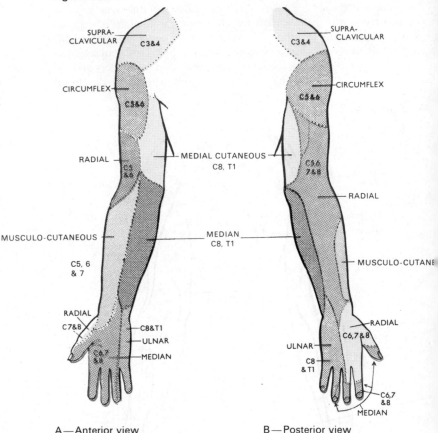

A—Anterior view   B—Posterior view

Distribution and origins of cutaneous nerves to the arm.

The thoracic nerves

The thoracic nerves do not intermingle to form a plexus. There are 12 pairs, the anterior primary rami of which pass between the ribs supplying the intercostal muscles and the overlying skin. They are known as the *intercostal nerves.*

The seventh to the twelfth thoracic nerves also supply the muscles and skin of the anterior abdominal wall.

The lumbar plexus

The lumbar plexus is formed by the anterior primary rami of the *first three* and *part of the fourth* lumbar nerves. The plexus is situated in front of the transverse processes of the lumbar vertebrae and behind the psoas muscle.

The main nerves formed by the lumbar plexus are the obturator and femoral nerves.

The obturator nerve supplies the adductor muscles and the skin on the medial aspect of the thigh and ends at the knee.

The femoral nerve is the largest branch of the lumbar plexus. It passes under cover of the inguinal ligament to supply the structures in the anterior and lateral aspect of the thigh and the anterior and medial aspect of the leg.

Figure 283

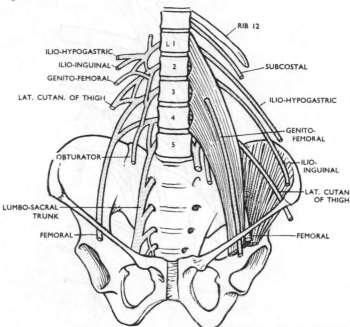

The lumbar plexus.

The sacral plexus

The sacral plexus is formed by the anterior primary rami of the *first, second and third sacral nerves* and by the lumbo-sacral trunk which is formed by the *fifth* and *part of the fourth lumbar nerves.* The sacral plexus lies in the posterior wall of the pelvic cavity.

The main nerve formed by the sacral plexus is the *sciatic nerve.*

The sciatic nerve is the largest nerve in the body and measures at its beginning approximately 2 cm ($\frac{3}{4}$ inch) in breadth. It passes through the greater sciatic foramen and descends through the posterior aspect of the thigh supplying the hamstring muscles. At the level of the middle of the femur it divides to form the medial and lateral popliteal nerves.

The medial popliteal nerve descends through the popliteal fossa to become the *posterior tibial nerve* which supplies the muscles and skin in the posterior aspect of the leg. Branches supply the sole of the foot and the toes.

Figure 284

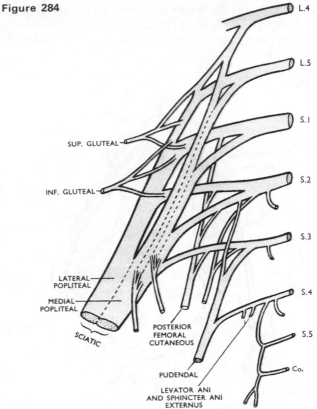

SUP. GLUTEAL

INF. GLUTEAL

LATERAL POPLITEAL

MEDIAL POPLITEAL

SCIATIC

POSTERIOR FEMORAL CUTANEOUS

PUDENDAL

LEVATOR ANI AND SPHINCTER ANI EXTERNUS

L.4

L.5

S.1

S.2

S.3

S.4

S.5

Co.

Formation of the sacral plexus.

The lateral popliteal nerve descends obliquely along the lateral aspect of the popliteal fossa to become the *anterior tibial nerve* which gives off the musculo-cutaneous nerve. These nerves supply the skin and muscles of the anterior aspect of the leg and the dorsum of the foot and toes.

The coccygeal plexus

The coccygeal plexus is a very small plexus formed by the anterior primary rami of the *fifth sacral and the coccygeal nerves.* The nerves from this plexus supply the skin in the area of the coccyx and the muscles of the pelvic floor.

Figure 285

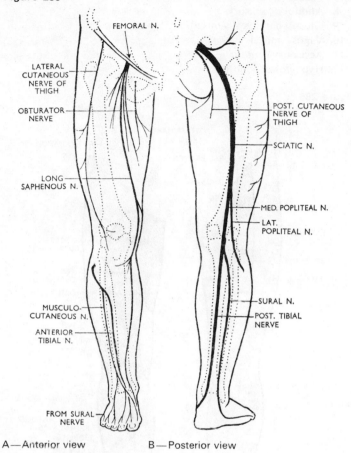

A—Anterior view   B—Posterior view

Main nerves of the leg

THE CRANIAL NERVES

There are 12 pairs of cranial nerves. These nerves arise directly from the brain. Some of the cranial nerves are sensory, some are motor and those containing both sensory and motor fibres are described as mixed. They are given names and numbers.

1. Olfactory—sensory
2. Optic—sensory
3. Oculomotor—motor
4. Trochlear—motor
5. Trigeminal—mixed
6. Abducent—motor

7. Facial—mixed
8. Auditory—sensory
9. Glossopharyngeal—mixed
10. Vagus—mixed
11. Accessory—motor
12. Hypoglossal—motor.

Figure 286

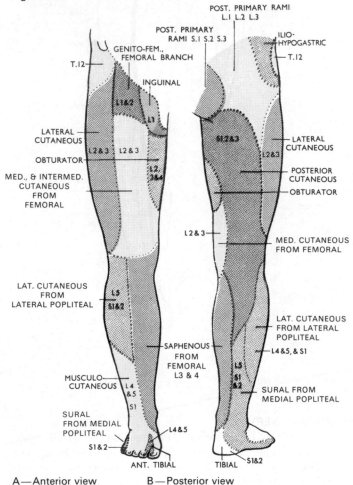

A—Anterior view   B—Posterior view

Distribution and origins of cutaneous nerves of the leg.

1. The olfactory nerves (sensory)

The olfactory nerves are the nerves of the sense of smell. Their sensory nerve endings and fibres arise in the upper part of the mucous membrane

Figure 287

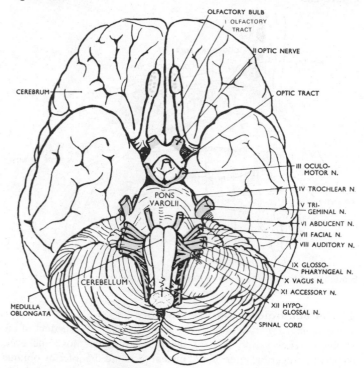

Diagram showing the position of the cranial nerves shown from below.

of the nose and pass upwards through the cribriform plate of the ethmoid bone. These nerves pass to the *olfactory bulb*, a group of nerve cells where synapses occur and the impulse is passed to a second neurone. The nerves then proceed backwards to the area for the perception of smell in the temporal lobe of the cerebrum.

2. The optic nerves (sensory)

The optic nerves are the nerves of the sense of sight. The fibres originate in the retinas of the eyes and they combine to form the optic nerves which are about 4 cm in length. They are directed backwards and medially through the posterior part of the orbital cavity. They then pass through the *optic foramina* of the sphenoid bone into the cranial cavity and join at the *optic chiasma* just above the pituitary gland. The nerves proceed backwards as the *optic tracts* to the *lateral geniculate body*. Impulses pass from these to the centre for sight in the occipital lobe of the cerebrum and to the cerebellum. In the occipital lobe sight is perceived, and in the cerebellum the impulses from the eyes contribute to the maintenance of balance.

Figure 288

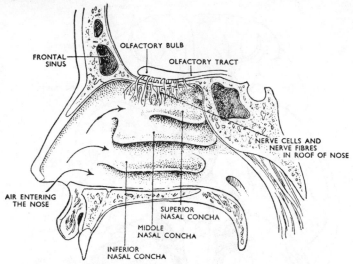

Branches of the olfactory nerve in the roof of the nose (diagrammatic).

3. The oculomotor nerves (motor)

The oculomotor nerves arise from nerve cells near the aqueduct of the midbrain. They supply four of the muscles which move the eyes, namely the superior, inferior and medial recti and the inferior oblique muscles. They also supply the ciliary muscles which control the focusing power of the eyes by altering the shape of the lens, the circular muscles of the iris producing constriction of the pupils and the muscles which control the movements of the upper eyelids.

4. The trochlear nerves (motor)

The trochlear nerves arise from a nucleus just behind the oculomotor nucleus. These nerves supply the superior oblique muscles of the eyes.

5. The trigeminal nerves (mixed)

The trigeminal nerves contain motor and sensory fibres and are among the largest of the cranial nerves. They are the chief sensory nerves for the face and head, receiving impulses of pain, temperature and pressure. The motor fibres stimulate the muscles of mastication.

The motor fibres arise from nerve cells which are situated just above the pons varolii and pass to the muscles of mastication.

There are three main branches of the trigeminal nerves each of which has motor and sensory fibres.

1. *The Ophthalmic Nerves*. These supply the lacrimal glands, the

Figure 289

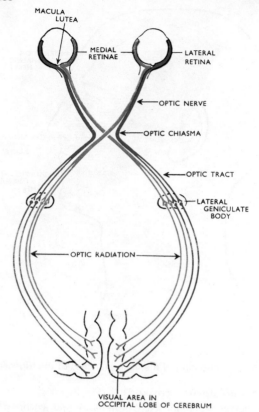

MACULA
LUTEA

MEDIAL
RETINAE

LATERAL
RETINA

OPTIC NERVE

OPTIC CHIASMA

OPTIC TRACT

LATERAL
GENICULATE
BODY

OPTIC RADIATION

VISUAL AREA IN
OCCIPITAL LOBE OF CEREBRUM

Illustration of pathway of optic nerves, showing the optic chiasma and optic tracts.

conjunctiva of the eyes, the forehead, eyelids, anterior aspect of the scalp and mucous membrane of the nose.

2. *The Maxillary Nerves.* These supply the cheeks, the upper gums, upper teeth and the lower eyelids.

3. *The Mandibular Nerves.* These are the largest of the three divisions of the trigeminal nerves. They supply the teeth and gums of the lower jaw, the pinna of the ears and the lower lip. Motor nerves from these branches supply the muscles of mastication.

6. The abducent nerves (motor)

The abducent nerves arise from a group of nerve cells lying beneath the floor of the fourth ventricle. They supply the lateral rectus muscles of the eyeballs

Figure 290

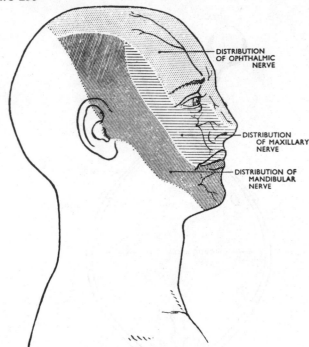

DISTRIBUTION OF OPHTHALMIC NERVE

DISTRIBUTION OF MAXILLARY NERVE

DISTRIBUTION OF MANDIBULAR NERVE

The peripheral distribution of the main branches of the trigeminal nerve.

7. The facial nerves (mixed)

The facial nerves are composed of both motor and sensory nerve fibres. The fibres arise from nerve cells in the lower part of the pons varolii. The motor fibres supply the muscles of facial expression. The sensory fibres convey impulses from the taste buds of the anterior two-thirds of the tongue to the taste perception area in the cerebral cortex.

8. The auditory nerves (sensory)

The auditory nerves are composed of two distinct sets of fibres:
    cochlear fibres
    vestibular fibres.

The cochlear fibres originate in the cochlea or inner ear and convey impulses to the hearing area in the cerebral cortex in both hemispheres where they are perceived as sound.

The vestibular fibres arise from the semi-circular canals of the inner ear and convey impulses to the cerebellum which are associated with the maintenance of equilibrium and balance.

9. The glossopharyngeal nerves (mixed)

The glossopharyngeal nerves are composed of motor and sensory fibres.

The fibres arise from nuclei in the medulla oblongata. The motor fibres stimulate the muscles of the pharynx and the secretory cells of the parotid glands.

The sensory fibres convey impulses from the posterior third of the tongue and from the tonsils and pharynx to the cerebral cortex

10. The vagus nerves (mixed)

The vagus nerves are composed of both motor and sensory fibres. They

Figure 291

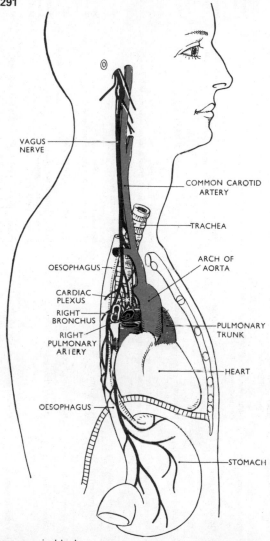

The vagus nerve, in black.

Summary of the Cranial Nerves

Name and No.	Central Origin	Distribution	Function
1. Olfactory (sensory)	Smell area in temporal lobe of cerebrum through olfactory bulb	Mucous membrane in roof of nose	Sense of smell
2. Optic (sensory)	Sight area in occipital lobe of cerebrum. Cerebellum	Retina of the eye	Sense of sight Balance
3. Oculomotor (motor)	Nerve cells near the floor of the aqueduct of the midbrain	Superior, inferior and medial rectus muscles, inferior oblique and ciliary muscles of the eye, and circular muscle fibres of the iris	Moving the eyeball, regulating the size of the pupils and focusing
4. Trochlear (motor)	Nerve cells near floor of aqueduct of midbrain	Superior oblique muscles of the eyes	Movement of the eyeball
5. Trigeminal (mixed)	Motor fibres from the pons varolii Sensory fibres from the trigeminal ganglion	Muscles of mastication Sensory to gums, cheeks, lower jaw, iris, cornea	Chewing Sensation from the face
6. Abducens (motor)	Floor of fourth ventricle	Lateral rectus muscles of the eye	Movement of the eye
7. Facial (mixed)	Pons varolii	Sensory fibres to the tongue Motor fibres to the muscles of the face	Sensation of taste Movements of facial expression
8. Auditory (sensory)	Hearing area of cerebrum	Organ of Corti in the cochlea	Sense of hearing
9. Glossopharyngeal (mixed)	Medulla oblongata	Back of tongue and pharynx Posterior third of tongue Parotid glands	Sense of taste Secretion of saliva Movements of pharynx
10. Vagus (mixed)	Medulla oblongata	Pharynx, larynx, lungs, heart, gall bladder, stomach Small and large intestine	Movement and secretion
11. Accessory (motor)	Medulla oblonagta	Sternomastoid, trapezius, laryngeal and pharyngeal muscles	Movement of the head and shoulders and pharynx and larynx
12. Hypoglossal (motor)	Medulla oblongata	Tongue	Movement of tongue

have a more extensive distribution than any of the other cranial nerves. They arise from nerve cells in the medulla oblongata and other nuclei, and pass through a foramen in the base of the skull downwards through the neck between the internal jugular vein and the internal carotid artery into the thorax and abdomen.

The motor fibres supply the muscles and secretory glands of the pharynx, larynx, trachea, heart, oesophagus, stomach, small intestine, pancreas, gall bladder, bile ducts, spleen, colon, kidneys and blood vessels in the thoracic and abdominal cavities.

The sensory fibres convey impulses from the mucous membranes of the larynx, trachea, lungs, oesophagus, stomach, small intestine and gall bladder to the brain.

11. The accessory nerves (motor)
The accessory nerves arise from nerve cells in the medulla oblongata and in the spinal cord. The fibres supply the sternomastoid and trapezius muscles. Branches join the vagus nerves and proceed to the pharyngeal and laryngeal muscles.

12. The hypoglossal nerves (motor)
The hypoglossal nerves arise from nerve cells in the medulla oblongata. They supply the muscles of the tongue and muscles surrounding the hyoid bone.

The Autonomic Nervous System
The autonomic or involuntary part of the nervous system controls the functions of the body which are carried out automatically. These functions do not reach consciousness under normal conditions.

The following list provides some examples of physiological activities controlled by the autonomic nervous system:
  the rate and force of the heart beat
  the calibre of the blood vessels
  the movements of the alimentary tract
  the secretion of the glands of the alimentary tract
  the contraction and relaxation of involuntary muscle
  the secretion from the sweat glands
  the size of the pupils of the eyes.

The *efferent nerves* of the autonomic nervous system arise from nerve cells in the brain. These emerge at various levels between the midbrain and the sacral region of the spinal cord. Many of them travel within the same sheath as the peripheral nerves of the central nervous system to reach the organs which they innervate.

For anatomical and physiological convenience the autonomic nervous system is described in two parts:
sympathetic and parasympathetic.

STRUCTURE OF THE SYMPATHETIC NERVOUS SYSTEM

Three neurones are involved in conveying impulses from their origin in the *hypothalamus* and the *medulla oblongata* to the effector organs and tissues.

1. The cell is in the brain and its fibre in the spinal cord.
2. The cell is in the lateral horn of grey matter in the spinal cord between the levels of the first thoracic and the fifth lumbar vertebrae. The fibre of this cell leaves the cord by the anterior root and terminates in a ganglion outside the spinal cord. This fibre is known as the *pre-ganglionic fibre*.
3. The cell is in one of the ganglia and its fibre conveys the impulse to the effector organ or tissue. This is the *post-ganglionic fibre* and it is generally longer than the pre-ganglionic fibre.

There are two types of arrangement of ganglia in the sympathetic part of the nervous system.

1. The *lateral chains of ganglia* which lie outside the neural canal on the lateral surfaces of the bodies of the vertebrae. The outflow of pre-ganglionic fibres occurs only between the first thoracic and the fifth lumbar vertebrae but the chains of ganglia extend upwards into the cervical region and downwards into the sacral region.
2. The *colateral ganglia*, consisting of the *coeliac* and the *superior* and *inferior mesenteric ganglia*, are made up of the cells from which the post-ganglionic fibres emerge.

The distribution of post-ganglionic sympathetic fibres can be seen in Figure 293.

STRUCTURE OF THE PARASYMPATHETIC NERVOUS SYSTEM

The parasympathetic part of the nervous system is sometimes referred to as the *cranio-sacral outflow* of the autonomic nervous system, because the nerves involved emerge mainly from the brain and the sacral region of the spinal cord. Two neurones are involved in the transmission of impulses from their source to the effector organ.

The cranial outflow

The pre-ganglionic fibres arise from nerve cells situated in the *midbrain, pons varolii and medulla oblongata*. Then travel towards ganglia situated in the walls of the organs to be innervated. Post-ganglionic fibres pass to the muscles or cells to be stimulated.

The *vagus nerves* are the most important nerves of the cranial outflow.

The sacral outflow

The nerves of the sacral outflow pass out with the second, third and fourth sacral nerves from the spinal cord. The preganglionic fibres

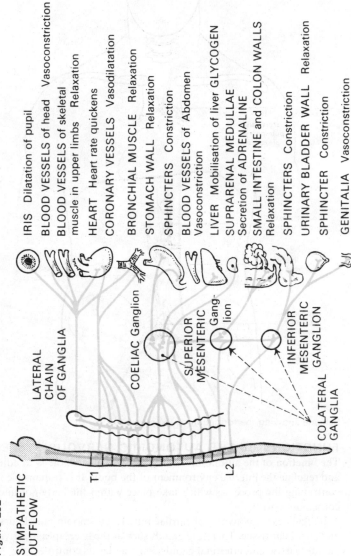

Figure 292

SYMPATHETIC OUTFLOW

IRIS Dilatation of pupil

BLOOD VESSELS of head Vasoconstriction

BLOOD VESSELS of skeletal muscle in upper limbs Relaxation

HEART Heart rate quickens

CORONARY VESSELS Vasodilatation

BRONCHIAL MUSCLE Relaxation

STOMACH WALL Relaxation

SPHINCTERS Constriction

BLOOD VESSELS of Abdomen Vasoconstriction

LIVER Mobilisation of liver GLYCOGEN

SUPRARENAL MEDULLAE Secretion of ADRENALINE

SMALL INTESTINE and COLON WALLS Relaxation

SPHINCTERS Constriction

URINARY BLADDER WALL Relaxation

SPHINCTER Constriction

GENITALIA Vasoconstriction

LATERAL CHAIN OF GANGLIA

COELIAC Ganglion

SUPERIOR MESENTERIC Ganglion

INFERIOR MESENTERIC GANGLION

COLATERAL GANGLIA

T1

L2

The sympathetic outflow.

travel with the pelvic nerves to the ganglia situated near or in the organs to be innervated. The post-ganglionic fibres then proceed to stimulate the organs.

Figure 293

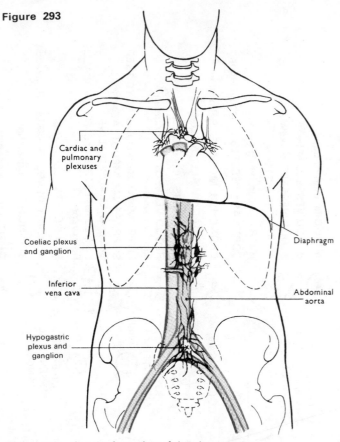

Scheme showing the formation of the plexuses.

FUNCTIONS OF THE AUTONOMIC NERVOUS SYSTEM

The function of the autonomic nervous system as a whole is to adjust and regulate the internal environment of the body. It is responsible for controlling the processes which take place within the body not under conscious control.

It regulates the activities of cardiac muscle, all smooth muscle tissue and glandular tissue. Therefore glands such as the liver, spleen, salivary glands, gastric and intestinal glands come under its control.

The majority of organs are innervated by *both* sympathetic and parasympathetic nerve fibres and the effect of one counteracts the effect of the other. In health they are very delicately balanced, for example, in the control of the size of the pupil of the eye.

Figure 294

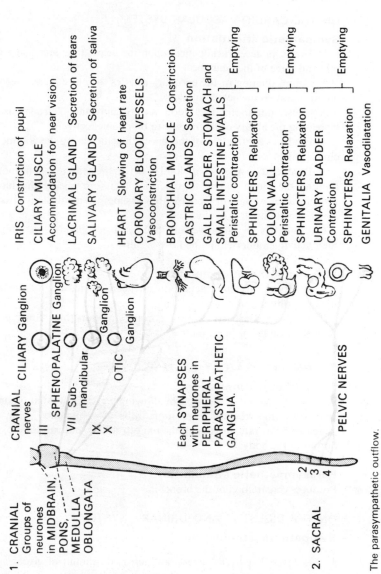

1. CRANIAL
Groups of
neurones
in MIDBRAIN,
PONS,
MEDULLA
OBLONGATA

CRANIAL   CILIARY Ganglion
nerves
III
        SPHENOPALATINE Ganglion
VII Sub-
     mandibular
            Ganglion
IX
X   OTIC
       Ganglion

Each SYNAPSES
with neurones in
PERIPHERAL
PARASYMPATHETIC
GANGLIA.

IRIS Constriction of pupil

CILIARY MUSCLE
Accommodation for near vision

LACRIMAL GLAND Secretion of tears

SALIVARY GLANDS Secretion of saliva

HEART Slowing of heart rate

CORONARY BLOOD VESSELS
Vasoconstriction

BRONCHIAL MUSCLE Constriction

GASTRIC GLANDS Secretion

GALL BLADDER, STOMACH and
SMALL INTESTINE WALLS
Peristaltic contraction   } Emptying

SPHINCTERS Relaxation

COLON WALL
Peristaltic contraction   } Emptying

SPHINCTERS Relaxation

URINARY BLADDER
Contraction   } Emptying

SPHINCTERS Relaxation

GENITALIA Vasodilatation

2. SACRAL

2
3
4

PELVIC NERVES

The parasympathetic outflow.

THE EFFECTS OF AUTONOMIC STIMULATION ON THE VARIOUS SYSTEMS OF THE BODY

ON THE CARDIO-VASCULAR SYSTEM

Sympathetic stimulation

1. Exerts an accelerating effect upon the heart, thus increasing the rate and force of the heart beat.

2. Causes dilation of the coronary arteries therefore increasing the blood supply to the heart muscle.

3. Causes dilation of the blood vessels supplying skeletal muscle, increasing the ability of the muscle to function effectively.

4. Produces contraction of the spleen thus increasing the volume of circulating blood.

5. Produces constriction of blood vessels in the periphery of the body, limiting the blood supply to the skin, therefore raising the blood pressure or providing for increased blood supply to highly active tissue, e.g., skeletal muscle.

6. Constricts the blood vessels in the salivary glands thus limiting their activity and lessening the flow of saliva.

Parasympathetic stimulation

1. Exerts an inhibitory action on the heart, decreasing its rate and force of contraction.

2. Produces constriction of the coronary arteries thus reducing the flow of blood to the heart muscle.

ON THE RESPIRATORY SYSTEM

Sympathetic stimulation

Produces dilatation of the bronchi allowing a greater amount of air to enter the lungs at each inspiration, thus in conjunction with the increased heart rate, increasing the oxygen intake and carbon dioxide output of the body.

Parasympathetic stimulation

Produces constriction of the bronchi.

ON THE DIGESTIVE AND URINARY SYSTEMS

Sympathetic stimulation

1. *The Liver.* The liver converts an increased amount of glycogen to glucose thus making more carbohydrate immediately available to provide energy.

2. *The Supra-renal Glands.* The supra-renal glands are stimulated to secrete adrenalin and noradrenalin which potentiates the effects of sympathetic stimulation.

3. *The Stomach and Small Intestine*. The activity of these two organs is reduced, thus the digestion and absorption of food is delayed.

4. *Urethral and Anal Sphincters*. Increases muscle tone which inhibits micturition and defaecation.

Parasympathetic stimulation

1. *The Stomach and Small Intestine*. The activity of these two organs is increased, thus speeding up digestion and absorption of food.

2. *The Pancreas*. The activity of both types of cell in the pancreas is increased, therefore there is an increased secretion of pancreatic juice and secretion of insulin.

3. *Urethral and Anal Sphincters*. Relaxation of the internal urethral sphincter is accompanied by contraction of the muscle of the bladder wall and micturition occurs.

Similarly relaxation of the internal anal sphincter is accompanied by contraction of the muscle of the rectum and defaecation occurs.

ON THE EYE

Sympathetic stimulation

This causes contraction of the radiating muscle fibres of the iris thus dilatation of the pupil occurs. Retraction of the eyelids takes place producing a look of alertness and excitement.

Parasympathetic stimulation

This causes contraction of the circular muscle fibres of the iris thus the pupil constricts. The eyelids tend to close producing the appearance of sleepiness.

Under normal healthy conditions there is a very fine balance between these two effects insuring an optimal degree of dilation or constriction of the pupils.

ON THE SKIN

Sympathetic stimulation

1. Causes greater activity of the sweat glands, therefore more sweat is produced and heat loss from the body is increased.

2. Produces contraction of the arrectores pilorum (the muscles in the skin). The effect of this is to produce the appearance of 'goose flesh'.

3. Causes constriction of the blood vessels.

SUMMARY OF FUNCTIONS OF AUTONOMIC NERVOUS SYSTEM

The functions of the involuntary nervous system therefore are to maintain a delicate balance within the body.

Sympathetic stimulation as a whole can be closely associated with the action of adrenalin and noradrenalin which are secreted by the medulla of the supra-renal glands. It prepares the body to deal with stressful situations, strengthening its defences in excitement, danger and the hazards of extremes of temperature. In other words the body is mobilised for 'fight or flight'.

Parasympathetic stimulation has a tendency to slow down body processes except the digestion and absorption of food. Its effect is that of a 'peace maker' allowing restoration processes to occur quietly and peacefully.

Normally the two systems function simultaneously producing a regular heart beat, normal temperature and an internal environment compatible with the immediate external surroundings.

15. The Special Senses

The Ear

The ear is the organ of hearing. It is supplied by the *auditory or 8th cranial nerve* which is stimulated by vibrations caused by sound waves.

Apart from the auricle or pinna the structures which form the ear are encased within the petrous portion of the temporal bone.

STRUCTURE

The ear is divided into three distinct parts:
the external ear
the middle ear or tympanic cavity
the inner ear.

THE EXTERNAL EAR

The external ear is divided, for descriptive purposes, into two parts:
the auricle or pinna
the external acoustic meatus.

The auricle

The auricle is the expanded portion which projects from the side of the head. It is composed of *fibro-elastic cartilage* covered with skin. The skin is covered with fine hairs and there are many sebaceous glands which open into the hair follicles. It is deeply grooved and ridged and the most prominent outer ridge is called the *helix*.

The lobule is the soft pliable part at the lower extremity and is composed of fibrous and adipose tissue richly supplied with blood capillaries and covered with skin.

The external acoustic meatus

This extends from the auricle to the *tympanic membrane* or ear drum. It consists of a cartilaginous portion and an osseous portion.

Figure 295

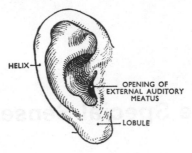

The lobe of the ear.

The cartilaginous portion forms the lateral one-third of the canal and the *osseous portion* the medial two-thirds. The meatus is approximately 4 cm ($1\frac{1}{2}$ inches) in length and is slightly 'S' shaped. It is important to remember this when examining or syringing the ear. To carry out a satisfactory examination or to syringe the canal properly, the auricle, in an adult, must be pulled gently backwards and upwards in order to straighten the canal.

The external acoustic meatus is lined with a continuation of the skin of the auricle and in the cartilaginous part there are modified sweat glands called *ceruminous glands* which secrete *cerumen* or wax. The skin lining the meatus extends over the outer part of the tympanic membrane.

Figure 296

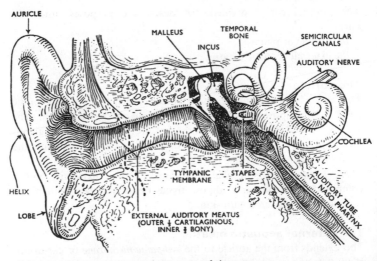

Scheme showing the general anatomy of the ear.

Figure 297

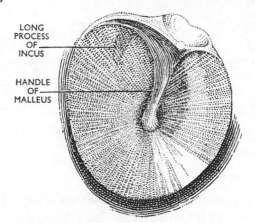

LONG
PROCESS
OF
INCUS

HANDLE
OF
MALLEUS

The tympanic membrane from the outside, showing the shadow of the malleus and part of the incus.

The tympanic membrane completely separates the external acoustic meatus from the middle ear. It is oval in shape being slightly broader at the top than the bottom. It is formed by three separate types of tissue.

1. The outer covering is *stratified epithelium*, a continuation of the lining of the external acoustic meatus.

2. The middle coat consists of *fibrous tissue*.

3. The inner lining consists of *cuboidal epithelium*, which is continuous with that of the middle ear.

THE MIDDLE EAR OR TYMPANIC CAVITY

The middle ear is an irregularly shaped cavity within the petrous portion of the temporal bone. It is lined with mucous membrane composed of cuboidal epithelium and filled with air. The air reaches the middle ear through a tube—*the auditory* or *pharyngotympanic tube.*

The auditory tube is approximately 4 cm (1½ inches) in length. It extends from the posterior wall of the naso-pharynx and passes upwards to open into the anterior part of the middle ear. It is lined with *ciliated epithelium* and allows for the passage of air between the pharynx and the middle ear. The presence of air ensures an equal pressure on both sides of the tympanic membrane.

The lateral wall of the middle ear is formed by the tympanic membrane.

The roof and floor are formed by the petrous portion of the temporal bone.

In the posterior wall there is an opening which leads into the *tympanic antrum*. This is a narrow cavity containing air which passes backwards into the *mastoid portion* of the temporal bone enabling air to pass into the mastoid air cells.

Figure 298

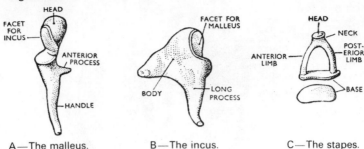

A—The malleus.   B—The incus.   C—The stapes.

The auditory ossicles.

The medial wall of the middle ear is composed of a thin wall of bone
in which there are two openings:
the fenestra vestibuli or oval window
the fenestra cochlea or round window.
The fenestra vestibuli is occupied by a small bone—*the stapes*. The
fenestra cochlea is filled with a fine sheet of *fibrous tissue*.

The auditory ossicles

Within the middle ear there are three minute bones known as the
auditory ossicles. The three bones form movable joints with one another
and extend across the cavity from the tympanic membrane to the
fenestra vestibuli.

The auditory ossicles are: *the malleous, the incus and the stapes.*

The Malleous. This lies in the lateral aspect of the middle ear. It is
shaped rather like a hammer, and presents *a head, neck and handle.*

The handle of the malleous is connected to the medial wall of the
tympanic membrane and the head forms a movable joint with the incus.

The Incus. This is the intermediate bone of the three. It has been
given its name because it is shaped like an anvil. It possesses a
body and *two processes*. It articulates freely with both the malleous and
the stapes.

The Stapes. This is the most medial of the ossicles, it is shaped like a
stirrup presenting *a head, neck, two limbs and a base.*

The head articulates with the incus and its base fits into the fenestra
vestibuli and is maintained in this position by an *annular ligament.*

The three ossicles are maintained in position by fine ligaments.

THE INNER EAR

The inner ear contains the organ of hearing and is generally described
in two parts.
The bony labyrinth
The membranous labyrinth.

The bony labyrinth

This is a cavity within the temporal bone. It is larger than the membranous labyrinth which fits into it, like a tube within a tube. The space between the bony walls and the membranous tube is filled with fluid known as *perilymph*. There is fluid also within the membranous tube and this is known as endolymph.

The part of the bony labyrinth nearest to the middle ear is expanded to form an area known as the *vestibule* which is separated from the middle ear by the fenestra cochlea and the fenestra vestibuli. Leading from the vestibule there are four separate and distinct canals. One is associated with hearing—*the cochlea;* the other three are associated with the maintenance of balance—*the semi-circular canals.*

The cochlea

The bony cochlea or labyrinth resembles a snail's shell and makes two and one half turns. It has a broad *base* near the vestibule and a narrower *apex* at the completion of its turns. The cochlea spirals round a central piece of bone known as the *modiolus*.

The membranous labyrinth lies within the bony labyrinth almost completely dividing the bony labyrinth into an area above called the *scala vestibuli* and an area below called the *scala tympani*.

Figure 299

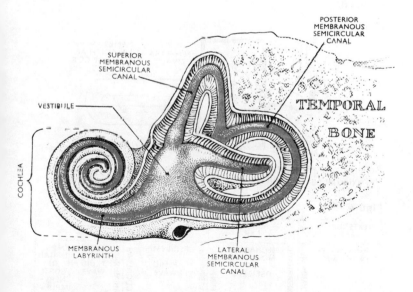

The inner ear showing the outer bony labyrinth, and the membranous labyrinth, inset in red.

The membranous labyrinth

The membranous labyrinth or membranous cochlea lies within the bony labyrinth being separated from the actual bone by the perilymph, and within it is the endolymph. Its walls are composed of a thin membrane. On its inferior aspect the membrane is known as the *basilar membrane* and on this lie special neuroepithelial cells and nerve fibres. Many of the cells are long and narrow and are arranged side by side. Some are surmounted by minute hair-like processes and are known as the *hair cells*. These cells and nerve fibres form the true organ of hearing known as the *organ of Corti*. The nerve fibres from the basilar membrane combine to form the *8th cranial or auditory nerve*, which passes through a foramen in the petrous portion of the temporal bone to reach the hearing area in the temporal lobe of the cerebral cortex of the cerebrum.

THE PHYSIOLOGY OF HEARING

Every sound that is produced causes vibrations or disturbances in the atmospheric air. These vibrations travel as a succession of waves known as *sound waves*.

The sound waves travel at approximately 0·2 miles per second or 1088 feet per second. It is the function of the ear to pick up these vibrations and direct them towards the cochlea, where the organ of Corti is adapted to transmit them, as nerve impulses, by the 8th cranial nerve to the *hearing area* of the cerebral cortex where they are perceived as sound.

Figure 300A

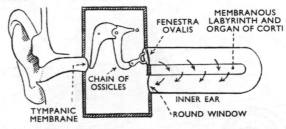

Schematic drawing of the ear showing the passage of vibrations from the external ear through the ossicles to the inner ear.

Figure 300B

EXTERNAL EAR			MIDDLE EAR			INNER EAR			
SOUND WAVES IN THE AIR	TYMPANIC	MEMBRANE	MECHANICAL MOVEMENT OF OSSICLES	FENESTRA	VESTIBULI	PERILYMPH FLUID WAVE	COCHLEAR	MEMBRANE	ENDOLYMPH FLUID → WAVE	ORGAN OF CORTI. NERVE IMPULSE

Illustration showing the conduction of sound waves from the external ear to the brain.

FUNCTION OF THE EXTERNAL EAR

The auricle is formed in such a way that the vibrations produced by any sound can be concentrated and direct through the external acoustic meatus towards the tympanic membrane. The tympanic membrane vibrates in harmony with the sound waves.

The cerumen secreted by the lining of the external acoustic meatus protects the canal and makes it waterproof.

FUNCTION OF THE MIDDLE EAR OR TYMPANIC CAVITY

The vibrations set up in the tympanic membrane are conveyed through the middle ear by the three auditory ossicles. The handle of the malleus is connected to the inner wall of the tympanic membrane and, as it vibrates, a corresponding movement of the malleus takes place. The movements of the malleus are transmitted via the incus to the stapes. The foot plate of the stapes fits accurately into the fenestra vestibuli. The vibrations transferred from the incus to the stapes causes the foot plate of the stapes to rock to and fro in the fenestra vestibuli. This carries the vibrations to the inner ear. The auditory ossicles would be unable to move and thus unable to transmit the vibrations unless air at atmospheric pressure was present in the middle ear. This air enters the middle ear from the naso-pharynx through the auditory tube.

FUNCTIONS OF THE INNER EAR

On the inner aspect of the fenestra vestibuli is the perilymph which

Figure 301

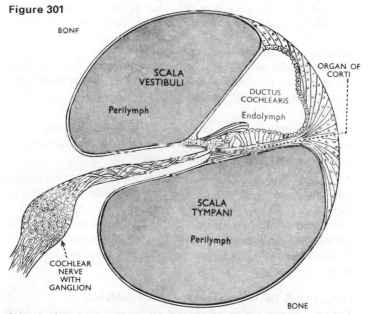

Scheme of the interior of the membranous labyrinth showing the organ of Corti.

surrounds the membranous labyrinth. The rocking movement of the stapes sets up wave motion in the perilymph which in turn causes vibration of the membranous labyrinth. The vibrations of the membranous labyrinth stimulate movement of the endolymph and subsequently the special neuro-epithelial cells forming the *organ of Corti* are stimulated. High pitched sounds stimulate the cells at the base of the cochlea, and low pitched sounds the cells at the apex.

These impulses are transmitted via the nerve fibres of the *cochlear nerve* which subsequently becomes the *8th cranial or auditory nerve*. The *auditory nerve* transmits the impulses to various nuclei in the pons varolii and midbrain. Some of the nerve fibres pass to the hearing area of the cerebral cortex on the same side and some cross over to the corresponding area in the opposite hemisphere. It is in the hearing area of the cerebral cortex that the nerve impulses are perceived as sound.

THE SEMI-CIRCULAR CANALS

It must be appreciated that the semi-circular canals have no auditory function. Anatomically they are closely associated with the inner ear but their functions are entirely different. Their functions are associated with equilibrium and balance.

ANATOMY OF THE SEMI-CIRCULAR CANALS

There are three semi-circular canals—the *superior, posterior* and *lateral*. They are described as lying in the three planes of space. They are situated above and behind the vestibule of the inner ear and open into it.

Structure of the semi-circular canals

The semi-circular canals like the cochlea are composed of an outer bony wall and inner membranous tubes. Within the membranous tubes there is endolymph and they are separated from the bony wall by perilymph.

The Utricle. Within the vestibule is a membranous sac, the utricle, which is roughly oblong in shape. The three membranous canals open into this sac at enlargements called the *ampullae.*

The Saccule. The saccule also lies within the vestibule. It is roughly cone-shaped and communicates by ducts with the utricle and the cochlea.

Within the walls of the utricle, saccule and ampullae there are fine epithelial cells with minute hair-like projections known as the *hair cells.* Amongst these hair cells are the minute nerve endings of the *vestibular nerve.*

FUNCTIONS OF THE SEMI-CIRCULAR CANALS

The semi-circular canals, utricle and saccule are associated with the position of the head in space. Any change of position of the head causes

Figure 302

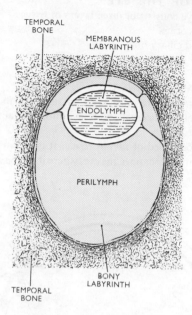

TEMPORAL
BONE

MEMBRANOUS
LABYRINTH

ENDOLYMPH

PERILYMPH

BONY
LABYRINTH

TEMPORAL
BONE

Scheme showing the interior of a semi-circular canal.

movement in the perilymph and endolymph. This movement stimulates the nerve endings and the hair cells in the utricle, saccule and ampullae. The impulses are then transmitted by the vestibular nerve, which joins the nerve from the cochlea of the inner ear, to form the *8th cranial or auditory nerve.* The vestibular branch of the 8th cranial nerve passes through an important relay station, the *vestibular nucleus,* and thence to the *cerebellum.*

Through its various connections the impulses transmitted by the vestibular nerve influence the movements of the eyes when the head is moved. Impulses also reach the cerebellum from muscles and joints, and thus the cerebellum is responsible for the maintenance of balance and equilibrium of the body as a whole.

The Eye

The eye is a special organ of the sense of sight. It is supplied by the *2nd cranial or optic nerve.*

The eye is situated in the orbital cavity, is spherical in shape and is approximately 2·5 cm (1 inch) in diameter. It is smaller in diameter than the orbital cavity and the intervening space is filled with fatty tissue. The bony walls of the orbit and the fat help to protect the eye from injury.

STRUCTURE OF THE EYE

The wall of the eye consists of three layers of tissue:
    the sclera and cornea
    the choroid and ciliary body
    the retina.

THE SCLERA AND CORNEA

The sclera forms the outermost layer of tissue of the posterior and lateral aspects of the eyeball and is continued anteriorly as the transparent *cornea*.

The sclera is composed of *fibrous tissue*. It is a firm membrane and maintains the form and shape of the eye, where it can be seen it forms the white of the eye.

Figure 303

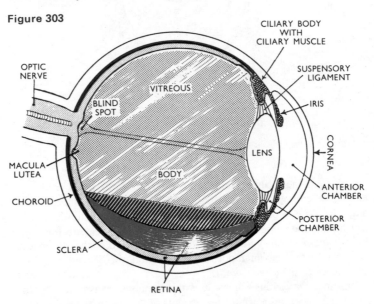

The eye, showing the cut edges of the different layers of tissue.

Anteriorly the sclera is replaced by the cornea which is a clear transparent membrane. It is composed of epithelial tissue and is *convex anteriorly*.

THE CHOROID

The choroid lines the inner surface of the sclera. It is very rich in blood capillaries and is deeply pigmented being chocolate brown in colour. It lines the posterior five-sixths of the sclera and is attached to it. Continuous with the choroid anteriorly is the *ciliary body*, which is a thickening of the choroid containing involuntary muscle fibres which form the *ciliary muscle*. This is one of the *intrinsic muscles* of the eye.

Two important structures are continuous with the ciliary body:
the iris
the lens.

The iris

This extends anteriorly from the ciliary body covering the anterior one-sixth of the eye, and lying just behind the cornea. It is a circular body composed of pigment cells and circular and radiating muscles fibres, thus forming the second *intrinsic muscle* of the eye. In the centre there is a circular aperture termed the *pupil*.

Figure 304

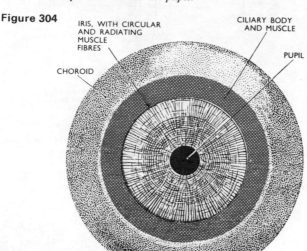

IRIS, WITH CIRCULAR
AND RADIATING
MUSCLE
FIBRES

CILIARY BODY
AND MUSCLE

PUPIL

CHOROID

Diagram showing the choroid, ciliary body and iris from the front.

The pupil varies in size depending upon the amount of light present. In a bright light the circular muscle fibres of the iris contract thus reducing the size of the aperture or *constricting the pupil*. In a dim light the radiating fibres contract, *dilating the pupil*.

The iris forms the coloured part of the eye. The colour depends upon the number of pigment cells present. In the albino, for example, there are no pigment cells present, and the number varies from a few cells in the deeper layer in light blue eyes to innumerable pigment cells throughout the whole structure in black eyes.

The lens

The lens is described as a circular bi-convex transparent body. It is suspended from the ciliary body by the *suspensory ligaments* and surrounded by a capsule.

The capsule and the lens itself are quite transparent and *highly elastic*. The lens lies immediately behind the pupil and its thickness is controlled by the ciliary muscle through its suspensory ligament.

Figure 305

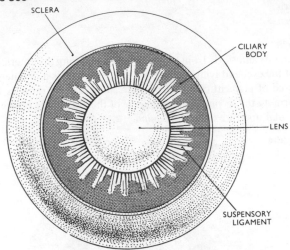

Diagram showing the position of the lens and suspensory ligament from in front, the iris has been removed.

Figure 306

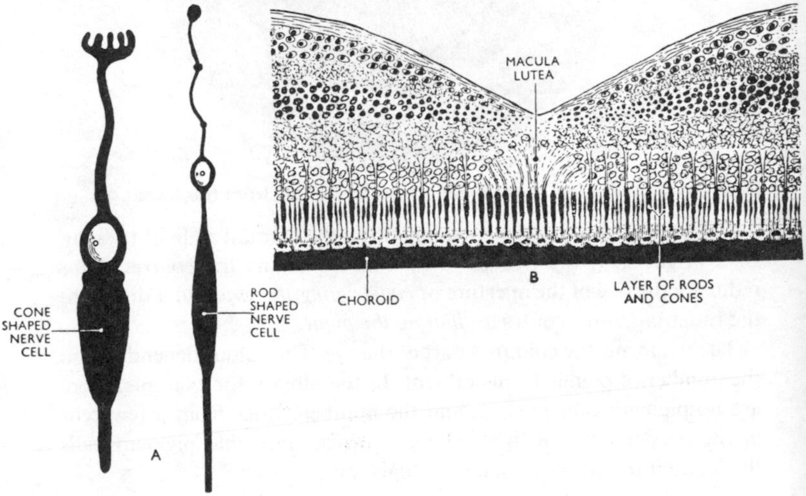

The retina.

A—An enlarged view of the rods and cones.
B—A highly magnified view of the retina, showing the macula lutea.

THE RETINA

The retina is the innermost layer of the wall of the eye. It is an extremely delicate membrane and is especially adapted to be stimulated by light rays. It is composed of several layers of nerve cells and nerve fibres. The most important and highly sensitive layer is the deepest layer nearest to the choroid which is known as the *layer of rods and cones*.

The retina lines about three-quarters of the eyeball and is thickest at the back and thins out anteriorly to end just behind the ciliary body. Near the centre of the posterior part of the retina there is a small depression. Here the cells appear yellow in colour, hence it is called the *macula lutea*. In the centre of the area there is a little depression called the *fovea centralis*. This is the most highly sensitive part of the retina consisting of only cone-shaped cells. Towards the anterior part of the retina there are fewer cones and more rod-shaped cells.

The retina has a purple tint due to the presence of *rhodopsin or visual purple* in the rods. This substance is bleached by bright light, and vitamin A is essential for its resynthesis.

Figure 307

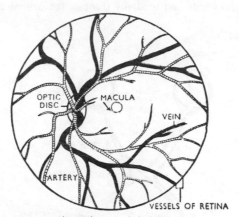

Interior of the eye as seen through an ophthalmoscope.

About 0·5 cm to the nasal side of the macula lutea all the fibres of the retina converge to form the *optic nerve* which passes backwards through the optic foramen of the sphenoid bone towards the cerebral cortex.

Where the optic nerve leaves the eye the area is termed the *optic disc*.

The optic disc is pierced by the central artery and vein of the retina. At this area there are no nerve cells therefore it is insensitive to light and termed the *blind spot*.

The Blood Supply. The arteries supplying the eyeball are the ciliary arteries which are branches of the ophthalmic arteries.

INTERIOR OF THE EYEBALL

Between the *cornea* and the *iris* there is a small space, *the anterior chamber* and between the *iris* and the *lens* there is another space, *the posterior chamber*. Both chambers contain a clear watery fluid known as the *aqueous humour*. This fluid filters into these chambers from capillaries in the ciliary body and is drained back into the circulation through the tiny ducts at the angle of the sclera and ciliary body.

Behind the lens and filling the cavity of the eyeball is the *vitreous body, or vitreous humour*. This is a soft, colourless, transparent jelly-like substance composed of 99 per cent water, some salts and mucoprotein. It maintains sufficient intraocular pressure to support the retina against the choroid and prevent the walls of the eyeball from collapsing.

THE OPTIC NERVES

The fibres of the optic nerve originate in the retina of the eye. All the fibres converge at a point about 0·5 cm to the nasal side of the macula lutea to form the optic nerve. The nerve pierces the choroid and sclera to pass backwards and medially through the orbital cavity. It then

Figure 308

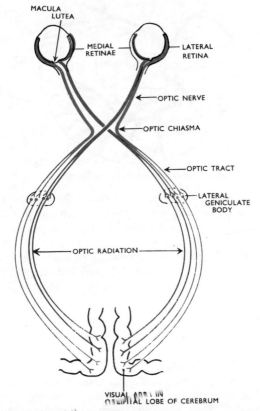

The optic nerves and their pathways.

passes through the optic foramen of the sphenoid bone, backwards and medially to meet its fellow from the other eye at the *optic chiasma*.

The optic chiasma

This is situated immediately in front of and above the pituitary gland which rests in the sella turcica of the sphenoid bone. In the optic chiasma the nerve fibres of the optic nerve from the *nasal side* of each retina *cross over* to the *opposite side*. The fibres from the *temporal side* of the retina *do not cross over* but continue backwards on *the same side*.

The optic tract

This is the term used to describe the optic pathway posterior to the optic chiasma. This consists of the nasal fibres from the retina of one eye and the temporal fibres of the retina from the other eye. The optic tracts pass backwards through the cerebrum to groups of nerve cells known as the *lateral geniculate bodies*.

The Lateral Geniculate Bodies. These lie just below and behind the thalamus. They consist of nerve cells and act as a relay station for the optic nerves. From here the nerve fibres proceed backwards and medially as the *optic radiations* to terminate in the *visual area* of the cerebral cortex in the occipital lobe of the cerebrum. Other neurones originating in the lateral geniculate body convey impulses from the eyes to the cerebellum where they contribute to the maintenance of balance.

THE PHYSIOLOGY OF SIGHT

Light waves travel at a speed of 186,000 miles per second. Light is reflected into the eyes from the various objects within the field of vision. White light is actually a combination of all the colours of the spectrum or rainbow, that is, red, orange, yellow, green, blue, indigo and violet. This can be demonstrated by passing white light through a glass prism which refracts or bends the rays of the different colours to a greater or

Figure 309

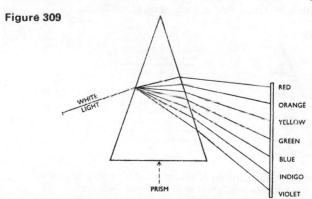

Drawing showing the breaking up of white light after passing through a prism.

lesser extent. The different colours have different wavelengths, therefore when white light is passed through a prism the different colours take different paths and fall upon objects at different levels.

Raindrops act as prisms and split sunlight into its constituent colours. The drops also act as tiny mirrors to reflect these separate colours into the eyes. When white light is broken up in this way the different colours form the *spectrum of visible light.*

The spectrum of light is extremely broad but only a small part of it is visible to the human eye. Violet is at the short wavelength end, and red at the long wavelength end of the visible spectrum. Beyond the short end there are the cosmic rays, X-rays and ultra-violet rays. Beyond the long end there are the infra-red heat waves, radar, and radio waves.

Figure 310

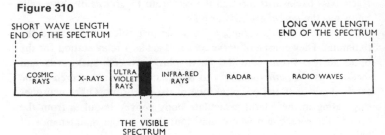

SHORT WAVE LENGTH
END OF THE SPECTRUM

LONG WAVE LENGTH
END OF THE SPECTRUM

| COSMIC RAYS | X-RAYS | ULTRA VIOLET RAYS | | INFRA-RED RAYS | RADAR | RADIO WAVES |

THE VISIBLE SPECTRUM

Drawing showing the spectrum of light.

A specific colour of the visible spectrum is perceived when that is the only wavelength of light which *is reflected* by an object and, when all other wavelengths are *absorbed.* For example, when an object appears *white* all wavelengths are being *reflected,* and when an object appears *black* all wavelengths are being *absorbed.* An article appears *red* when only the *red wavelength is reflected* and *all others are absorbed.*

In order to achieve clear, uniocular vision light reflected from objects within the visual field is focused on to the retina.

Figure 311

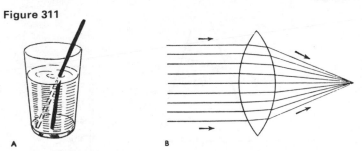

A   B

Drawing showing the bending or refraction of light rays when they pass from a less dense medium to a more dense medium.

A—Through water.
B—Through a biconvex lens.

Two processes are involved in producing a clear image on the retina or photo-sensitive area of the eye:

refraction of the light rays

accommodation of the eyes.

REFRACTION OF THE LIGHT RAYS

When light rays pass from a medium of one density to a medium of a different density they are refracted or bent. This principle is used in the eye to focus light on the retina.

Several structures which are more dense than air are involved in the refraction of light in the eye.

Before reaching the retina light rays pass successively through the cornea, the aqueous humor, the lens and the vitreous body.

Figure 312

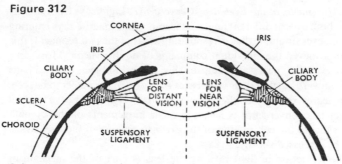

Scheme showing the change of shape of the lens in the eye when focusing light.

The lens is the only structure within the eye which can change its refractory ability. Light rays from objects 20 feet or more away from the eyes travel in parallel lines and require relatively little refraction by the lens. As objects move nearer to the eyes the rays become more divergent and therefore require to be refracted more acutely.

The lens is a bi-convex, elastic transparent body suspended by the suspensory ligament from the ciliary muscle. When light rays enter the eye from near objects the function of the lens is to refract these rays on to the macula lutea. To do this the *ciliary muscle contracts forwards* releasing the pull on the suspensory ligament. Due to the presence of elastic fibres the lens now becomes shorter and thicker increasing its convexity and therefore its refracting power.

It will be appreciated that looking at near objects will 'tire' the eyes more quickly due to the continuous use of the ciliary muscle.

ACCOMMODATION OF THE EYES TO LIGHT

Three factors are involved in accommodation:

the pupils

the movement of the eyeballs, called convergence

the lens.

The Size of the Pupils. This influences accommodation by controlling the amount of light entering the eye. *In a bright light the pupils are constricted. In a dim light they are dilated.*

If the pupils were dilated in a bright light too much light would enter the eye and damage the retina. In a dim light if the pupils were constricted, insufficient light would enter the eye to stimulate the nerve endings in the retina and nothing would be seen.

The iris consists of a layer of circular and a layer of radiating smooth muscle fibres. Contraction of the circular fibres constricts the pupil, and contraction of the radiating fibres dilates it. The size of the pupil is controlled by nerves of the autonomic nervous system. *Sympathetic stimulation dilates the pupils. Parasympathetic stimulation constricts the pupils.*

The Movements of the Eyeballs—Convergence. Light rays from objects enter both eyes at different angles. It is important that the rays entering the two eyes stimulate the corresponding areas of the two retinae. If this is not achieved the individual complains of double vision. Movement of the eyeballs is necessary to ensure this. For near objects both eyeballs move inwards so that their visual axes *converge* on the object to be seen. The nearer the object the greater the convergence. For a distant object less convergence is necessary. The movements of the eyeballs occur due to the contraction of the *extrinsic muscles* of the eyes which will be discussed shortly.

The refraction of light rays by the lens is part of the mechanism of accommodation.

Figure 313

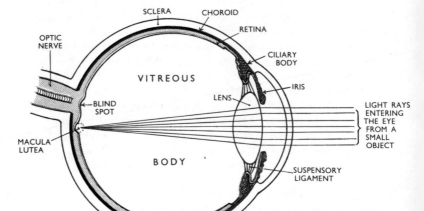

The pathway of light rays through the eye, focused on the macula lutea.

FUNCTION OF THE RETINA

The retina is described as the *photosensitive* mechanism of the eye. The light sensitive cells in the retina are the *rods and cones*.

The rods are more sensitive than the cones. They are stimulated by *low intensity or dim light*, for example, by the dim light in the interior of a cinema.

The cones are sensitive to *bright light and colour*. The effect of different wavelengths of light on the cones results in the perception of different colours. In a bright light the light rays must be focused on the macula lutea, in this area there is an abundance of cones and a photo-chemical change within these cells gives rise to a nerve impulse which is transmitted to the optic nerve. In a dim light the rods are stimulated.

Figure 314

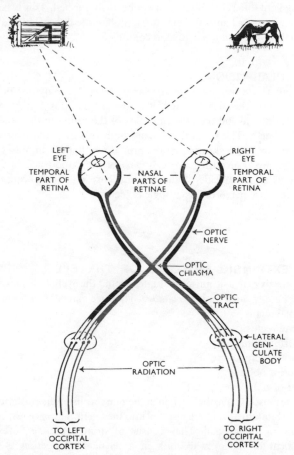

Diagram illustrating binocular vision, showing both objects stimulating both retinae.

The rods are more prevalent towards the periphery of the retina. Present in the rods is a pigmented protein, *the visual purple or rhodopsin*, which is very sensitive to a low intensity of light. This is bleached in bright light (therefore the rods cannot be stimulated by bright light). Soon after going into a dim light the visual purple is reformed and the rods can be stimulated. '*Dark adaptation*' has occurred, in other words, an individual can see in the '*dark*' *when he* '*gets used to it.*' Vitamin A is essential for the reconstitution of the visual purple.

The reader may have noticed that it is easier to see a star in the sky at night if the head is turned slightly away from it. When this happens the dim rays from the star stimulate the area of the retina where there is the greatest concentration of rods. If looked at directly the light rays are not of sufficient intensity to stimulate the less sensitive cones in the area of the macula lutea. It may also have been noticed that in dim evening light different colours can not be distinguished. This is because the low intensity light rays can not stimulate the cones which are the only cells in the retina sensitive to colour.

BINOCULAR VISION

Both retinae are stimulated by the same object, but only one object is seen because the light is reflected on to the nasal part of the retina in one eye and the temporal part of the retina in the other eye, and vice versa. The nasal fibres from both retinae cross over at the optic chiasma and the temporal fibres do not, thus impulses from both eyes are transmitted through each of the optic tracts and optic radiations to either side of the cerebral cortex in the occipital lobe where they are perceived as a single object.

THE EXTRINSIC MUSCLES OF THE EYE

There are six extrinsic muscles situated round the eyeball. Four of these muscles are *straight* and two *oblique*. They are named as follows:

   medial rectus
   lateral rectus
   superior rectus
   inferior rectus
   superior oblique
   inferior oblique.

These muscles have their origin in the bony walls of the posterior part of the orbital cavity and are inserted into the sclera. They are responsible for moving the eyeball. They consist of striated muscle fibres. The movement of the eyes to look in a particular direction is under voluntary control but the co-ordination of movement of the eyes during accommodation to near or distant vision is controlled by reflex action.

Figure 315

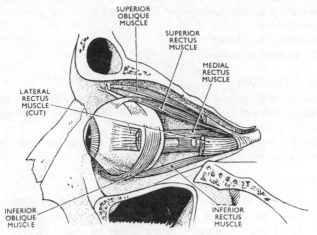

The extrinsic muscles of the eye.

The medial rectus rotates the eyeball *inwards.*
The lateral rectus rotates the eyeball *outwards.*
The superior rectus rotates the eyeball *upwards.*
The inferior rectus rotates the eyeball *downwards.*

Usually when the lateral rectus of one eyeball contracts the medial rectus of the other eyeball contracts—unless the individual consciously 'squints'.

The superior oblique rotates the eyeball so that the cornea turns in a *downwards and outwards* direction.

The inferior oblique rotates the eyeball so that the cornea turns *upwards and outwards.*

It is to be appreciated that rarely do any of the extrinsic muscles of the eyeball contract singly. The free, quick movements of the eyeball occur due to the contraction of more than one muscle at a time.

Nerve supply
The oculomotor or 3rd cranial supplies:
    superior rectus
    inferior rectus
    medial rectus
    inferior oblique
    iris
    ciliary muscle.

The trochlear or 4th cranial nerve supplies the superior oblique.

The abducens or 6th cranial nerve supplies the lateral rectus.

THE ACCESSORY ORGANS OF THE EYE

The eye is a delicate organ which is protected by several structures:

the eyebrows

the eyelids and lashes

the lacrimal apparatus.

THE EYEBROWS

These are two arched eminences of skin surmounting the supra-orbital margins of the frontal bone. Numerous hairs project obliquely from the surface of the skin.

The function of the eyebrows is to protect the anterior aspect of the eyeball from sweat and dust.

THE EYELIDS AND EYELASHES

The eyelids are two movable folds situated above and below the front of the eye, on their free edges are strong outgrowths of hair—the eyelashes. From without inwards the eyelids are composed of:

a thin covering of skin

a thin sheet of areolar tissue

two muscle plates—the *orbicularis oculi and levator palpebrae superioris*

a thin sheet of dense connective tissue—the *tarsal plate*—which supports the other structures and maintains the shape and form of the upper lids.

a delicate lining of conjunctiva.

The Conjunctiva. This is a fine transparent membrane which *lines the eyelids* and is *reflected over the front of the eyeball.* When the eyelids

Figure 316

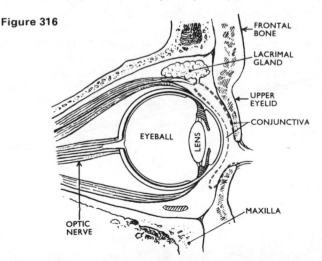

Diagram showing the position of the conjunctiva represented by the red dotted line.

are closed this becomes a closed sac. It protects the delicate cornea and front of the eye. When a drug is put into the eye it is placed in the lower conjunctival sac.

The function of the eyelids and eyelashes is to protect the eye from injury. If injury is feared or the conjunctiva touched very lightly the eyelids close. This is termed the conjunctival or corneal reflex.

When the *orbicularis oculi* contracts the eyes close. When the *levator palpabrae* contract the eyelids open. Where the upper and lower lids of the eye come together at the corners these areas are known as the medial and lateral *canthi* of the eye.

Figure 317

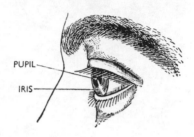

PUPIL

IRIS

Drawing showing the protection to the eye provided by the eyebrows and eyelashes.

THE LACRIMAL APPARATUS

This consists of:

    the two lacrimal glands and their ducts
    the lacrimal canaliculi
    the lacrimal sac
    the naso-lacrimal duct.

The lacrimal glands are situated in recesses in the frontal bones on the lateral aspect of each eye just behind the supra-orbital margin. Each gland is approximately the size and shape of an almond, and is composed of *secretory epithelial cells* which form alveoli. The glands secrete the *tears* which are composed of water, salts and the enzyme lysozyme which has a bactericidal action.

The tears leave the glands by several small ducts and pass over the front of the eyes and under the lids towards the medial canthus. At the medial canthus of the eye the tears drain into the two *lacrimal canaliculi*. One canaliculus lies above the other separated by the *caruncle*. The caruncle is a small red body situated at the medial canthus. The tears then drain into the *lacrimal sac*.

The lacrimal sac is situated in a fossa in the lacrimal bone and is the upper expanded end of the *naso-lacrimal duct*.

The naso-lacrimal duct is a membranous canal approximately 2 cm in length. It extends from the lower part of the lacrimal sac to open into the nasal cavity, at the level of the inferior concha.

Figure 318

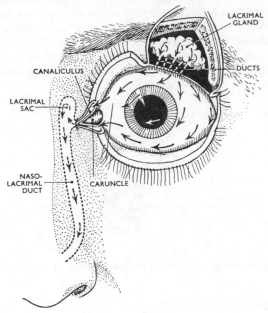

The lacrimal apparatus, the arrows show the flow of the tears across the eye.

The tears which have passed over the anterior aspect of the eyeball flow into the lacrimal canaliculi, to the lacrimal sac and through the naso-lacrimal duct into the nasal cavity.

The tears are secreted by the lacrimal glands continuously and their function is to bathe the front of the eyeball, washing away any dust, grit and micro-organisms. Due to the presence of the enzyme lysozyme, micro-organisms present on the front of the eyeball are destroyed. In emotional states the secretion of tears may be increased and, if the naso-lacrimal duct cannot convey them all into the nasal cavity, they overflow.

The Nose and the Sense of Smell

The anatomy of the nose was discussed in the chapter dealing with the respiratory system. The nose, therefore, is an organ with a dual function:

an organ of the respiratory system
an organ of special sense.

The *olfactory nerves or the 1st cranial nerves* are the sensory nerves of smell. They have their origins in the mucous membrane in the *olfactory region* of the nasal cavity.

The olfactory region lies in the roof of the nose and includes the superior nasal concha. Within the mucous membrane lining of the roof of the nose and covering the superior nasal concha there are minute nerve endings and nerve fibres. The nerve fibres are gathered into bundles then pass through the *cribriform plate of the ethmoid bone* to the *olfactory bulb*. In the *olfactory bulb* inter-connections and synapses occur. From the bulb, bundles of nerve fibres form the *olfactory tracts* which pass backwards to the olfactory area in the *temporal lobes* of the cerebral cortex in each hemisphere where the impulses are interpreted.

PHYSIOLOGY OF SMELL

The sense of smell in human beings is generally less acute than in other animals.

Figure 319

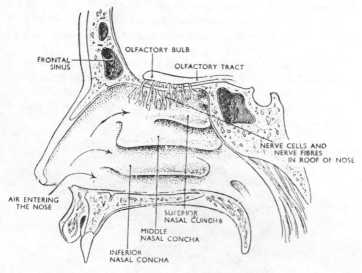

Scheme showing the nerv cells and fibres of the olfactory nerve in the roof of the nose.

All odorous materials give off particles of their substance and these chemical particles are carried into the nose with the inhaled air. They dissolve in the secretions of the mucous membrane and will only stimulate the nerve cells of the olfactory region when in solution.

Figure 320

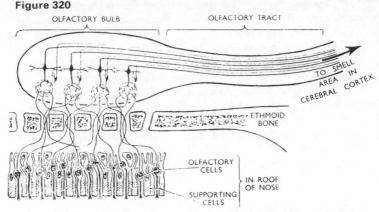

A magnified impression of the nerve cells and fibres in the roof of the nose, also showing the olfactory bulb and olfactory nerve tract.

When an individual is continuously exposed to an odour, perception of the odour decreases and eventually ceases. The loss of perception only affects the specific odour. This adaptation probably occurs both in the cerebrum and in the nerve endings in the nose.

This modification of perception may be dangerous when it is associated with the escape of unpleasant smelling poisonous gases, e.g., coal gas.

The air entering the nose is heated and convection currents carry the inspired air to the roof of the nose—'sniffing' transmits the currents of air more quickly to the roof of the nose therefore delicate odours are appreciated more fully. The sense of smell may improve the appetite, if the odours are pleasant.

Figure 321

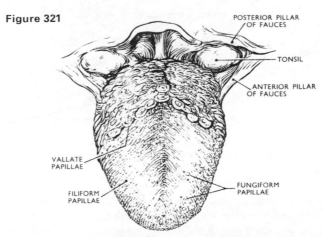

The tongue showing the papillae.

The Tongue and the Sense of Taste

The anatomy of the tongue was described in the chapter on the digestive system. *The tongue is the organ of taste* and is supplied by the *7th cranial (facial) nerve and the 9th cranial (glossopharyngeal) nerve.*

The tongue is a voluntary muscle covered with stratified squamous epithelium. The surface presents many tiny projections which are known as *papillae*. They vary in shape and size being smaller at the front than at the back.

The vallate or circumvallate papillae are the largest and present a V-shaped row at the back of the tongue. Each appears like a tower with a moat surrounding it.

The fungiform papillae, so-called because of their resemblance to a toadstool or fungus, are distributed mainly on the sides and tip of the tongue.

The filiform papillae are long and slender and are found mainly on the anterior two-thirds of the tongue.

Embedded within the papillae are small collections of slender cells arranged in bundles. These cells are the *taste buds*. Each cell has a minute branch from the nerves of taste. The taste buds open on to the surface of the papillae at a tiny *pore* and the cells collected together show fine hair-like projections.

Taste buds are also present in the soft palate, pharynx and epiglottis.

Figure 322

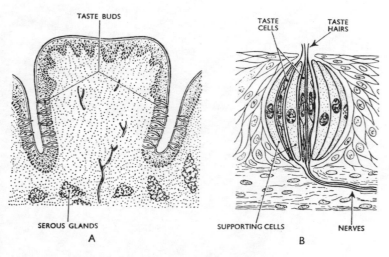

A magnified view of the taste buds in the tongue.

A—Showing a circumvallate papilla.
B—Showing a taste bud.

PHYSIOLOGY OF TASTE

There are four fundamental sensations of taste, *sweet, sour, bitter and salt*. The large variety of other 'tastes' which human beings experience are either a combination of two or more of the fundamental tastes, or are associated with the sense of smell.

The tongue is also sensitive to *temperature, pain, touch and pressure*. Highly spiced substances are appreciated due to temperature and touch rather than taste.

Taste is a chemical sensation, therefore substances must be dissolved either in water or saliva before they can be appreciated. The substances in solution enter the pores of the taste buds and stimulate the hair-like endings of the cells. The impulses are then transmitted to the thalamus and then to the *taste area* in the cerebral cortex in either hemisphere where the impulses are interpreted.

The taste buds are not evenly distributed over the whole area of the tongue, and it is thought that appreciation of the four tastes take place at different parts of the tongue.

Sweet and salty tastes are appreciated mainly at the *tip of the tongue*.

Sour tastes are perceived mainly at the *sides of the tongue*.

Bitter tastes are recognised most strongly at the *back of the tongue*.

The dorsum of the tongue is scarcely sensitive to taste, although some physiologists consider it to be sensitive to salty tastes.

The tip of the tongue is highly sensitive to *pain, temperature, touch and pressure*.

16. The Endocrine System or System of Ductless Glands

The endocrine system consists of glands widely separated from one another and possessing no direct anatomical relationship. The glands are commonly referred to as the *ductless glands* because the secretions which they produce *do not leave* the gland through a duct or a canal but pass *directly* from the cells into the blood stream.

The secretions produced by the endocrine glands are called *hormones*.

A hormone can be described as a chemical substance which, having been formed in one particular organ or gland is carried in the blood stream to another organ, probably quite far distant, where it has its effect, influencing its function, growth and nutrition.

The internal environment of the body is controlled partly by the autonomic nervous system and partly by the endocrine glands. The hormones secreted by the ductless glands are mainly excitatory in nature stimulating activity in other organs.

In a study of anatomy and physiology it is not usual to introduce a discussion of changes occurring during disease or signs and symptoms which may appear in individuals due to abnormal functioning of organs, but in the study of the ductless glands it may help the student to obtain a clearer picture of the normal functions if a short summary is given of the changes which can occur when these glands are not functioning satisfactorily. The main abnormalities which occur in association with the ductless glands are caused by an over secretion (hypersecretion), or an under secretion (hyposecretion), of hormone.

The endocrine system consists of the following glands:
1 pituitary gland
1 thyroid gland
4 parathyroid glands
2 adrenal or suprarenal glands
the islets of Langerhans in the pancreas
1 thymus
1 pineal gland or body

2 ovaries in the female

2 testes in the male

The *ovaries* and *testes* secrete hormones closely associated with the reproductive systems therefore their functions will be understood more readily if studied in conjunction with chapter 17.

The Pituitary Gland

The pituitary gland is reddish-grey in colour and roughly oval in shape. It is situated in the *sella turcica or hypophyseal fossa* of the sphenoid bone, and is approximately 1·5 cm long and 0·5 cm in diameter.

The gland is attached to the brain by a stalk which is continuous with the part of the brain known as the hypothalamus. Immediately above and slightly in front of the gland lies the optic chiasma. There is communication between the hypothalamus and the pituitary gland by means of nerve fibres and a complex of blood vessels.

Figure 323

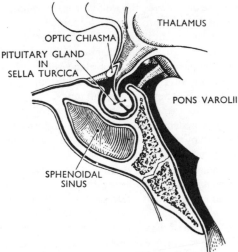

Diagram showing the position of the pituitary gland.

STRUCTURE

The pituitary gland is described as

having three parts or lobes:

the anterior lobe or pars anterior

the middle lobe or pars intermedia

the posterior lobe or pars nervosa or neurohypophysis.

The Anterior Lobe. This is composed of *secretory epithelial cells* which differ in shape and size. The majority of the cells contain granules and they can be distinguished one from the other.

The Middle Lobe. This is composed of *secretory epithelial cells* of varying shape and size, also containing granules.

Figure 324

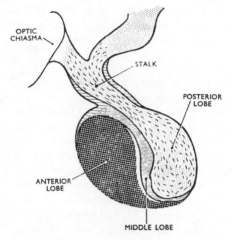

Diagram showing the lobes of the pituitary gland.

The Posterior Lobe. This is composed of specialised secretory cells termed *pituicytes*, and contains minute nerve fibres which arise from the nerve cells in the hypothalamus. *The hypothalamus* is a mass of nerve cells situated in the midbrain and the stalk of the pituitary gland is a continuation of this structure. The pituitary gland has a close association with the hypothalamus both through its nerve and blood supply.

The blood supply to the pituitary is by the superior and inferior hypophyseal arteries which are branches from the internal carotid artery.

Venous drainage is into one of the venous sinuses of the brain.

FUNCTIONS

The pituitary gland is often described as the '*master gland*' of the endocrine system because it secretes a number of hormones which regulate and control the activity of other endocrine glands. The functions of the anterior and posterior lobe will be described separately.

FUNCTIONS OF THE ANTERIOR LOBE

The anterior lobe of the pituitary gland secretes several hormones:

the growth hormone

the thyrotrophic hormone or thyroid stimulating hormone (TSH)

the adrenocorticotrophic hormone (ACTH).

the lactogenic hormone (prolactin).

the gonadotrophic hormones which include the follicle stimulating hormone (FSH) and the luteinizing hormone (LH).

The secretion of *parathyrotrophic, diabetogenic* and *pancreatrophic hormones* is doubtful. It may be that one or other of the hormones mentioned above has an effect on the islets of Langerhans in the pancreas, the parathyroid glands and on the utilisation of glucose and fat by the cells of the body.

Functions of the growth hormone

This hormone stimulates growth directly and in conjunction with other hormones. It affects the growth in length of the long bones by promoting the growth of the epiphyseal cartilage. Growth of long bones stops when ossification overtakes the growth of this cartilage.

Growth hormone promotes protein anabolism, the absorption of calcium from the bowel and the conversion of glycogen to glucose.

Cells in the hypothalamus are believed to control the secretion of the growth hormone. It is in highest concentration in the blood until the individual has reached his full stature, but a continual supply is necessary to stimulate the repair and replacement of body tissue throughout life.

Functions of the thyrotrophic hormone (TSH)

This hormone controls the growth and activity of the *thyroid gland*. It influences the uptake of iodine, the synthesis of thyroxin and triiodothyronine by the thyroid gland and the release of stored hormones into the blood stream.

The amount of thyrotrophic hormone secreted varies from time to time and its secretion is influenced by the amount of thyroid hormones circulating in the blood at any given time. If the concentration of these hormones in the blood falls the anterior lobe of the pituitary secretes more of the thyrotrophic hormone and this stimulates the thyroid gland into activity. On the other hand if the blood level of thyroxine and triiodothyronine is high the pituitary gland decreases its output of TSH. The hypothalamus has a stimulating effect on the production of thyrotrophic hormone especially by increasing the secretion when the body is cold and decreasing secretion when it is hot.

Functions of the adrenocorticotrophic hormone (ACTH)

This hormone stimulates the *cortex* of the *adrenal gland* to produce its hormones. The amount of ACTH secreted depends upon the concentration in the blood of the hormones from the adrenal cortex and on stimulation of the pituitary gland by the hypothalamus. This latter influence on the production of ACTH is particularly important during emotional and physical stress.

Functions of the lactogenic hormone (prolactin)

The lactogenic hormone has a direct effect upon the *breasts or mammary glands* immediately after the delivery of a baby and the expulsion of the placenta (after-birth). In conjunction with other hormones it stimulates the breasts to secrete milk.

During the months of pregnancy the breasts develop in preparation for lactation. This development is *not due* to prolactin, but to the activity of the *ovarian hormones*. It is only after the delivery of the baby and the expulsion of the placenta that the lactogenic hormone is secreted which stimulates the secretory cells to produce milk. During pregnancy the ovarian hormones inhibit the secretion of prolactin. Also involved in the functions of the breast in the postpartum period is a hormone *oxytocin* secreted from the *posterior lobe* of the pituitary gland. *Oxytocin* amongst other functions stimulates the *flow* of milk through the ducts of the breast.

Figure 325

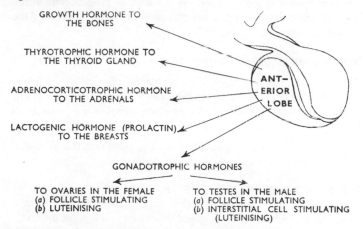

GROWTH HORMONE TO
THE BONES

THYROTROPHIC HORMONE TO
THE THYROID GLAND

ADRENOCORTICOTROPHIC HORMONE
TO THE ADRENALS

ANT–
ERIOR
LOBE

LACTOGENIC HORMONE (PROLACTIN)
TO THE BREASTS

GONADOTROPHIC HORMONES

TO OVARIES IN THE FEMALE
(a) FOLLICLE STIMULATING
(b) LUTEINISING

TO TESTES IN THE MALE
(a) FOLLICLE STIMULATING
(b) INTERSTITIAL CELL STIMULATING
(LUTEINISING)

Scheme showing the hormones secreted by the anterior lobe of the pituitary gland.

Three main processes are associated with lactation:

1. *The growth and development of the breasts* during pregnancy influenced mainly by *ovarian hormones*.

2. *The secretion of milk* in which a group of hormones, the galactopoetic complex, is involved. *The galactopoietic complex* is believed to include prolactin, the growth hormone, the adreno-corticotrophic hormone and the thyrotrophic hormone.

3. *The flow of milk* from the breast is only possible when *oxytocin*, secreted by the posterior lobe of the pituitary is present. Sucking plays an essential part in reflex stimulation of the pituitary gland. When sucking stops lactation will continue only for a short time.

Functions of the gonadotrophic hormones

The anterior lobe secretes *two gonadotrophic or sex hormones* in both the female and the male:

the follicle stimulating hormone (FSH)

the luteinising hormone, sometimes called the *interstitial cell stimulating hormone* in the male (LH).

Female gonadotrophic hormones

The Follicle Stimulating Hormone. The target organs are the ovaries where FSH stimulates the development and ripening of the ovarian follicle (see page 429). During its development the ovarian follicle secretes its own hormone *oestrogen.* As the level of oestrogen increases in the blood so the follicle stimulating hormone secretion is reduced.

The Luteinising Hormone. This hormone promotes the final maturation of the ovarian follicle, ovulation (discharge of the mature ovum) and the formation of the *corpus luteum* which secretes the second ovarian hormone known as *progesterone.* As the level of progesterone in the blood increases there is a gradual reduction in the production of the luteinising hormone.

Male gonadotrophic hormones

The Follicle Stimulating Hormone. This has an effect on the epithelial tissue of the *seminiferous tubules* in the testes. Under the influence of this hormone the *seminiferous tubules* produce *spermatozoa* (male germ cells).

The Interstitial Cell Stimulating Hormone. This stimulates the *interstitial cells* in the testes to secrete the hormone *testosterone.*

THE FUNCTIONS OF THE POSTERIOR LOBE

The posterior lobe and the hypothalamus of the cerebrum form a functional unit which controls several important functions in the body. There are several theories regarding the secretion of the posterior lobe, one is that the gland actually *forms the hormone,* and its *secretion* is controlled by the nerve fibres which pass to the lobe from the hypothalamus. The secretion is known as *pituitrin* which contains two hormones:

oxytocin or pitocin

antidiuretic hormone (ADH) or vasopressin.

Oxytocin

Oxytocin promotes contraction of the uterine muscle and contraction of the myoepithelial cells of the lactating breast, squeezing milk into the large ducts behind the nipple. In late pregnancy the uterus becomes very sensitive to oxytocin. The amount secreted during labour is increased.

Figure 326

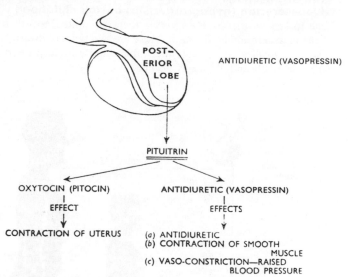

Scheme showing the hormones secreted by the posterior lobe of the pituitary gland.

Antidiuretic hormone (ADH) or vasopressin

This has a number of functions.

1. *Antidiuretic Effect.* This hormone affects the tubules of the kidney controlling the reabsorption of water. There are osmo-receptors in the hypothalamus which respond to changes in the osmotic pressure of the blood. When the osmotic pressure is high the osmo-receptors respond by stimulating the secretion of antidiuretic hormone, which increases the reabsorption of water by the renal tubules. When the osmotic pressure of the blood is low the secretion of the antidiuretic hormone is inhibited by the hypothalamus and the urine flow is increased. In this way the water balance of the body is maintained.

   If the antidiuretic hormone is lacking reabsorption of water is reduced and large amounts of dilute urine are excreted. This condition is called *diabetes insipidus.*

2. *Effect on Involuntary Muscle.* Under the influence of the anti-diuretic hormone the smooth muscle of the intestine, gall bladder, urinary bladder and blood vessels contracts. Contraction of the muscle layer in the blood vessel walls raises the blood pressure.

THE FUNCTIONS OF THE MIDDLE LOBE

There is no definite knowledge that this part of the pituitary secretes a hormone although it may be associated with skin pigmentation.

ABNORMALITIES OF PITUITARY ACTIVITY
Hyposecretion (hypopituitarism) during childhood
Due to lack of the *growth hormone* there is stunted growth of the skeleton with resultant *dwarfism*. There are two types of dwarf.

The Lorain Type. A minute individual, fairly well proportioned who is mentally and physically normal.

Figure 327

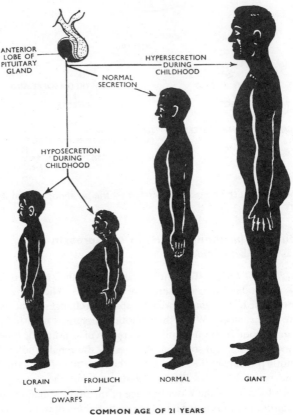

ANTERIOR
LOBE OF
PITUITARY
GLAND

HYPERSECRETION
— DURING —
CHILDHOOD

NORMAL
SECRETION

HYPOSECRETION
DURING
CHILDHOOD

LORAIN   FROHLICH   NORMAL   GIANT

DWARFS

COMMON AGE OF 21 YEARS

Scheme showing the effects of the growth hormone on the skeleton.

Frohlich's Type. In this type of dwarfism growth is stunted, there is considerable obesity, sexual development is incomplete and there is mental deficiency.

Hyposecretion (hypopituitarism) during adult life
This condition is fortunately very rare and is probably due to atrophy of the anterior lobe of the gland. The condition is manifested by early *senility*. It appears to effect females more than males. The skin becomes

dry and wrinkled, the hair grey and sparse, the body emaciated, and there is atrophy of the sexual organs with cessation of menstruation in the female. This condition is known as *Simmond's disease.*

Hypersecretion or hyperpituitarism during childhood

Due to increased secretion of the growth hormone *gigantism* occurs. There is excessive skeletal growth and the individual may be 8 or 9 feet tall.

Figure 328

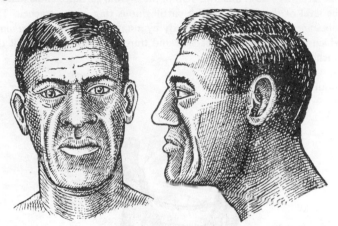

EXCESSIVE GROWTH OF BONES OF FACE
AND THICKENING OF SKIN

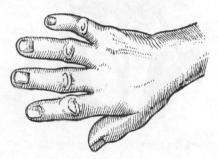

SPADE LIKE HANDS

Diagram illustrating the effect of hypersecretion from the anterior lobe of the pituitary gland in adult life.

Hypersecretion or hyperpituitarism during adult life

This results in a condition known as *acromegaly.* There is excessive growth of the bones of the face, specially the frontal bone and the mandible. The hands and feet become large and spade-like. There is also thickening of the skin on the face and hands. There are frequently signs of over secretion of the trophic hormones from the anterior lobe affecting their target glands.

The Thyroid Gland

The thyroid gland is situated in the neck in association with the larynx and trachea at the level of the fifth, sixth and seventh cervical and first thoracic vertebrae. It is a highly vascular gland, brownish red in colour and surrounded by a fibrous capsule. It consists of *two lobes* which lie one on either side of the lower half of the thyroid cartilage and extend to the level of the fourth or fifth cartilaginous ring of the trachea. The lobes are joined together anteriorly by a narrow portion, *the isthmus,* which lies in front of the second and third cartilaginous rings of the trachea.

Figure 329

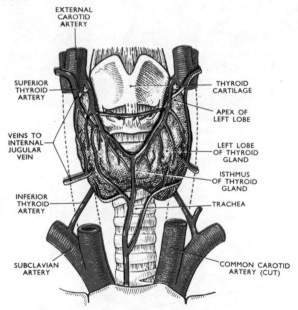

Diagram illustrating the position of the thyroid gland.

The lobes are roughly conical in shape. *The apex* is the upper narrow part, and the lower broader part, the *base*. The lobes are approximately 5 cm (2 inches) long and 3 cm ($1\frac{1}{4}$ inches) wide. The isthmus is about 1·25 cm ($\frac{1}{2}$ inch) long and 1·25 cm wide.

The arterial blood supply to the gland is through the superior and inferior thyroid arteries. The superior thyroid artery is a branch of the external carotid artery. The inferior thyroid artery is a branch of the subclavian artery.

The venous return is by the thyroid veins which drain into the internal jugular vein.

STRUCTURE

The gland is composed of *epithelial cells* which vary in shape depending upon the activity of the gland. They vary from *cubical in quiescent periods* to *columnar in active conditions*. The cells are arranged in a single layer in a roughly spherical or oval fashion round a central space, *the alveolus*. This space contains a thick viscid fluid known as *colloid*. The cells are in direct contact with supporting connective tissue and blood capillaries.

Figure 330

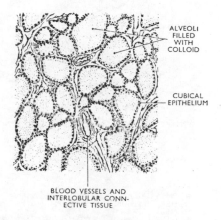

ALVEOLI FILLED WITH COLLOID

CUBICAL EPITHELIUM

BLOOD VESSELS AND INTERLOBULAR CONNECTIVE TISSUE

The minute structure of the thyroid gland.

FUNCTION

Iodine present in the blood is taken up by the gland and the hormones *thyroxine* and *triiodothyronine* are formed. These hormones are stored in the gland in the form of thyroglobin and are released into the blood stream as required. Most of the *iodine* ingested by the body is either utilised in the formation of thyroxine and triiodothyronine or stored in the gland for future use. The uptake of iodine from the blood, the formation of the hormones and their release into the blood is stimulated by the thyrotropic hormone from the anterior lobe of the pituitary gland.

The effects of thyroxine and triiodothyronine on other body functions are widespread.

1. An adequate supply is essential for normal, mental and physical development.

2. They are responsible for the maintenance of healthy skin and hair.

3. They are associated with nerve stability, probably controlling the excitability of the nerve fibres.

4. They control the utilisation of oxygen in the body. In this way they have a controlling influence on the basal metabolic rate.

5. They stimulate the absorption of carbohydrate from the small intestine.

6. They influence heat production during the katabolism of nutrient materials in the cells.

Figure 331

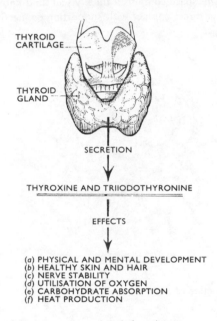

The thyroid gland, its hormones and their functions.

ABNORMALITIES OF THYROID ACTIVITY
Hyposecretion (hypothyroidism) in infancy
This results in the development of the condition known as *cretinism*.

Characteristics of Cretinism

1. Lack of skeletal development resulting in stunted body growth.

2. Lack of satisfactory development of the nervous system and because of this the child is mentally defective. The tongue appears too large and protrudes from a sagging mouth.

3. The features are coarse, the face expressionless and the lips are thickened.

4. The pulse and respiration rates are slow and the temperature is subnormal.

5. The skin is thick and dry and the hair is brittle and sparse.

6. The abdomen ptrotrudes.

7. There is general sluggishness of all body processes as a result of the low basal metabolic rate.

Figure 332

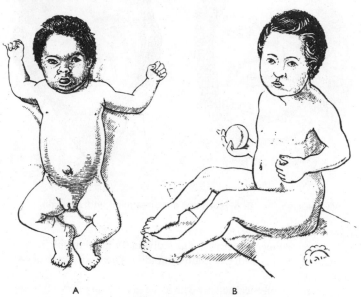

A   B

A. A cretin.   B. A cretin after treatment with thyroxin.

Hyposecretion (hypothyroidism) in adult life

Hyposecretion in adult life is responsible for the condition known as *myxoedema*.

Characteristics of Myxoedema

1. Slowing of all mental activity. The individual is slow in speech and thought. Mental effort cannot be maintained and thinking is difficult, giving the appearance of stupidity.

2. Slowing of all physical activity. Every movement is slow and lethargic and frequent prolonged rests are taken even after slight exertion.

3. The skin becomes coarse, dry and scaly with the hair becoming dry and brittle and falling out.

4. There is considerable general oedema.

5. The body temperature is subnormal and the individual 'feels the cold'.

6. The basal metabolic rate is reduced.

Hypersecretion (hyperthyroidism) or thyrotoxicosis

This condition generally occurs in *adult life*.

Figure 333

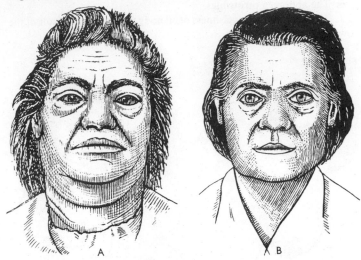

A. Myxoedema.   B. Myxoedema after treatment with thyroxin.

Characteristics of hyperthyroidism

1. Increased mental activity. The individual's mind jumps from one thought to another with tremendous rapidity. There is extreme nervous excitability.

2. Increased physical activity with a fine involuntary nervous tremor. Voluntary movements may be rapid and uncontrolled.

3. The body temperature is above normal. The pulse rate is rapid, from 100 to 160 beats per minute with a corresponding increase in respiration rate.

4. The skin is moist and, due to dilation of the blood vessels in the dermis, the sweat glands are over active. The palms of the hands are always 'sweaty' and the hair is damp, lank and greasy in appearance.

5. There is considerable loss of weight even though the appetite is increased.

6. The basal metabolic rate is greatly increased and the individual appears to be literally 'burning himself up'.

7. There may or may not be *exophthalmos*, that is, protrusion or forward displacement of the eyes. The cause of this protrusion of the eyeballs is not completely understood, but it is thought to be due to increased post-orbital pressure and possible weakening of the extrinsic muscles of the eye which is the result of excess thyrotrophic hormone.

Endemic or simple goitre

Iodine is an essential requirement for the formation of thyroxine. In certain countries such as Switzerland, the Himalayan regions and in the Pyrenees, where the iodine content of water and soil is low or absent,

Figure 334

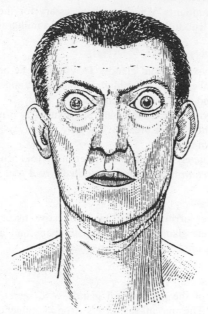

Hyperthyroidism with exophthalmos.

simple goitre is fairly prevalent. This enlargement is believed to be due to overstimulation of the gland by the thyrotrophic hormone from the anterior lobe of the pituitary. The actual swelling is the result of the accumulation of colloid.

Figure 335

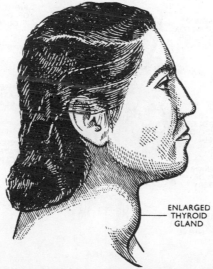

ENLARGED
THYROID
GLAND

Illustration showing the enlarged thyroid gland in simple goitre.

Only small amounts of iodine are necessary for the formation of thyroxine and triiodothyronine; a fraction of a milligram per day.

The Parathyroid Glands

There are four parathyroid glands. They usually lie on the posterior aspect of the lobes of the thyroid gland and are arranged in two pairs. The *superior* and the *inferior parathyroids*.

The parathyroids are quite small, approximately 6 mm ($\frac{1}{4}$ inch) in length and slightly less in breadth. They are roughly oval in shape and yellowish-brown in colour.

The arterial supply is through the inferior thyroid artery.
The venous return is through the middle thyroid vein.

STRUCTURE

The glands are surrounded by fine connective tissue capsules. The cells forming the glands are spherical in shape and are arranged in columns. They are known as the *principal cells*.

FUNCTIONS

The function of the parathyroid glands is to secrete the hormone *parathormone*. The amount secreted is strongly influenced by the level of calcium in the blood.

Functions of parathormone

The calcium level in blood plasma is approximately 4·5 to 5·5 mEq/litre (9 to 11 mg per cent). The constant level of calcium in the plasma is

Figure 336

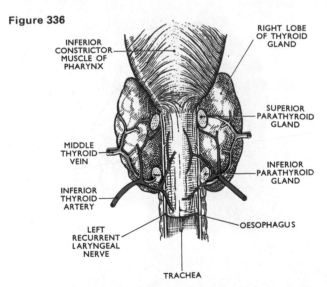

INFERIOR
CONSTRICTOR
MUSCLE OF
PHARYNX

RIGHT LOBE
OF THYROID
GLAND

SUPERIOR
PARATHYROID
GLAND

MIDDLE
THYROID
VEIN

INFERIOR
PARATHYROID
GLAND

INFERIOR
THYROID
ARTERY

LEFT
RECURRENT
LARYNGEAL
NERVE

OESOPHAGUS

TRACHEA

Diagram showing the position of the parathyroid glands.

maintained by the hormone parathormone. The calcium must be in satisfactory amounts and in a suitable chemical form to activate nervous and muscular tissue and to be deposited in bone.

ABNORMALITIES OF PARATHYROID ACTIVITY
Hyposecretion (hypoparathyroidism)
If there is a decrease in the secretion of parathormone an upset in calcium metabolism occurs. The usable calcium in the blood is reduced and *hypocalcaemia* develops. It is thought that, in this condition, much of the calcium in the blood is bound to other substances so that the *level of free calcium* in the blood is reduced. As a result the muscles go into a type of spasm known as *tetany*.

Characteristics of Tetany
   1. Muscle twitching leading to actual muscle spasm particularly of the hands and feet termed *carpo-pedal spasm*.
   2. There is twitching of the face and spasm of the eye muscles.
   3. In infants and children spasm of the muscles of the larynx occurs accompanied by cyanosis. After a pause without breathing a sharp inspiration occurs accompanied by a high-pitched 'crowing' sound.

Figure 337

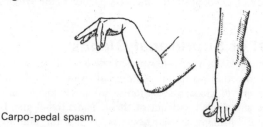

Carpo-pedal spasm.

Hypoparathyroidism occurs only very rarely and may be due to:
   1. removal of the parathyroid glands in error during the operation of thyroidectomy.
   2. removal of the glands if the thyroid gland is malignant.
   3. some local condition of the glands themselves when they are unable to secrete parathormone.

Hyperparathyroidism
Hyperparathyroidism is rare and is probably due to a tumour of one or more of the parathyroid glands.

   If there is an increase in the secretion of parathormone the calcium level of the blood is increased and this is termed *hypercalcaemia*. Many experiments have been carried out to explain this phenomenon but no certain conclusions have been formed. It may be that parathormone controls the activity of the osteoclasts—the cells which destroy bone.

If there is a hypersecretion of parathormone then the osteoclasts are over active, there is an excessive destruction of bone and non-utilisation of calcium resulting in the softening of the bones and a high blood calcium. In the areas of destruction in the bones fibrous cysts develop. This condition is called *osteitis fibrosa cystica*.

Figure 338

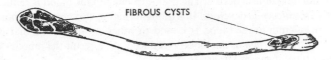

FIBROUS CYSTS

Diagram showing the presence of fibrous cysts in bone due to hyperparathyroidism.

Characteristics of Osteitis Fibrosa Cystica

1. The bones become painful and soft, and fractures occur frequently.

2. There is weakness of skeletal muscle.

3. There is an increase in urinary calcium with a tendency to the formation of renal calculi (stones in the kidney) with subsequent damage to the kidneys.

4. There may be calcium deposits in other tissues such as the arteries and lungs.

The Adrenal or Suprarenal Glands

There are two adrenal glands which are situated one on each side of the vertebral column on the posterior abdominal wall behind the peritoneum. They are in close association with the kidneys lying immediately above and slightly in front of them. The adrenal glands are surrounded by a capsule of areolar tissue containing fat, and are enclosed within the renal fascia but separated from the kidneys by areolar tissue.

The right adrenal gland is roughly pyramidal in shape, the left roughly semilunar in shape, and both are yellowish-brown in colour. Each gland is approximately 4 cm (1½ inches) long and 3 cm (1¼ inches) thick.

The arterial blood supply to the glands is by branches from the abdominal aorta and renal arteries.

The venous return is through the suprarenal vein which leaves the hilum of the gland and drains into the inferior vena cava from the right gland and into the renal vein from the left gland.

STRUCTURE

The glands are composed of two distinct parts which differ both anatomically and physiologically. The outer part is known as the *cortex* and the inner part as the *medulla*.

Figure 339

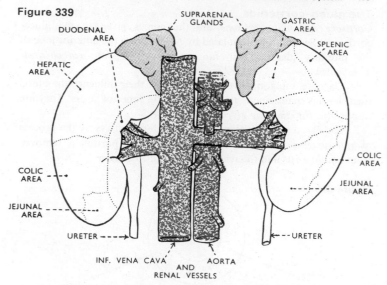

DUODENAL AREA

SUPRARENAL GLANDS

GASTRIC AREA

SPLENIC AREA

HEPATIC AREA

COLIC AREA

JEJUNAL AREA

URETER

COLIC AREA

JEJUNAL AREA

URETER

INF. VENA CAVA   AORTA
AND
RENAL VESSELS

Diagram showing the position of the adrenal or suprarenal glands.

CORTEX

The cortex is yellowish in colour and completely surrounds the medulla. It is composed of three layers of cells.

The outer layer consisting of large cells is called the *zona glomerulosa*.

The middle layer consisting of columns of cells is called the *zona fasciculata*.

The inner layer where the cells become interlaced into a network is called the *zona reticularis*.

MEDULLA

The medulla is completely surrounded by the cortex and comprises only about one-tenth of the gland. It is reddish in colour. The cells of the medulla are arranged in columns with many intermingling blood capillaries. The medulla is composed of sympathetic ganglion cells which have no post-ganglionic fibres. They are stimulated by pre-ganglionic fibres to produce the hormones *adrenalin* and *noradrenalin*.

FUNCTIONS

The *cortex* and the *medulla*, although an anatomical entity, have entirely different functions.

THE FUNCTIONS OF THE CORTEX

The cortex produces three groups of hormones:

The *gluco-corticoids* produced by the three zones.

The *mineralo-corticoid* produced by the zona glomerulosa only.

The *sex hormones* produced by the three zones.

The gluco-corticoids

Cortisone and hydrocortisone are the names given to the gluco-corticoids. Secretion is stimulated by ACTH from the anterior lobe of the pituitary gland. Their function is to *regulate carbohydrate metabolism.*

1. They are antagonistic to insulin. Under their influence the blood sugar level is maintained, and even raised in times of stress. They are responsible for *changing glycogen to glucose.*

2. The blood sugar level may be raised during stress by the process of *gluconeogenesis.* In this case the nitrogenous portion is removed from amino acids and the residue converted to glucose.

Figure 340

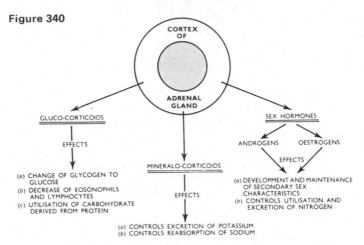

Scheme showing hormones secreted by the adrenal cortex.

3. These hormones tend to decrease the number of eosinophils and lymphocytes in the blood, and to increase the neutrophil count.

The mineralo-corticoids

Aldosterone is the name given to the main physiological mineralo-corticoid. Its functions are associated with the maintenance of the electrolyte balance in the body.

Aldosterone stimulates the reabsorption of sodium by the renal tubules and when the amount of sodium *reabsorbed* is increased the amount of potassium *excreted* is increased. Indirectly this affects water excretion as the amount of water excreted is related to the amount of electrolyte excreted.

The amount of aldosterone produced is influenced by the sodium level in the blood. If there is a fall in the sodium blood level more aldosterone is secreted and more sodium reabsorbed.

Angiotensin produced by the kidneys stimulates the secretion of aldosterone.

The sex hormones

Female—*oestrogens*

Male—*androgens*.

The secretion of these hormones by the adrenal cortex is controlled by ACTH.

Their functions are:

1. To influence, both in the male and the female, the development and maintenance of the secondary sex characteristics.

2. To increase the deposition of protein in muscles, and reduce the excretion of nitrogen in the male.

ABNORMALITIES OF CORTICAL ACTIVITY

Hyposecretion

Hyposecretion of hormones from the adrenal cortex results in the development of the condition known as *Addison's disease*.

Characteristics of Addison's Disease

1. Loss of appetite.
2. Muscular weakness.
3. Loss of weight due to loss of water,
4. Inability to maintain the normal deposition of protein in the muscles.
5. Subnormal temperature and reduced metabolic rate.
5. Hypoglycaemia.
6. Increased blood potassium and decreased blood sodium.
7. Pigmentation or bronzing of the skin especially the exposed parts, for example, the hands and face.

Hypersecretion of the adrenal cortical hormones in childhood

This condition predisposes to precociousness, that is, too early development of the sexual organs and secondary sex characteristics. Precocious children also have a tendency to show unusual muscular development and obesity. Children suffering from these characteristics have sometimes been described as 'infant Hercules'.

Hypersecretion from the adrenal cortex in adults

This results in a condition described as *Cushing's Syndrome* which is characterised by *virilism* in the female, that is, a tendency to develop male characteristics. Hair may grow excessively on the chest and pubic region with an increase and darkening of the facial hair. There will be amenorrhoea (cessation of the menstrual flow), and gradual atrophy of the breasts.

In the male *feminism* may develop, that is, a tendency to develop female sex characteristics.

Other abnormalities which may be present include:

wasting of the muscles due to excess break down of protein
muscular weakness due to potassium loss through the kidneys
hyperglycaemia and glycosuria due to the upset in carbohydrate
metabolism.

Figure 341

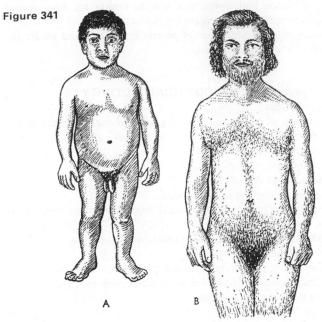

A

B

A. Hypersecretion from the adrenal cortex in childhood, showing
overdeveloped male child.

B. Hypersecretion from the adrenal cortex in an adult. Showing the
masculine distribution of hair in a female.

THE FUNCTIONS OF THE MEDULLA

The medulla of the adrenal gland is an extension of the sympathetic
part of the autonomic part of the nervous system. It produces two
hormones, *adrenalin* and *noradrenalin* when stimulated by pre-
ganglionic sympathetic nerve fibres.

These hormones have the same effects on the body as sympathetic
stimulation.

1. Dilatation of the coronary arteries, thus increasing the blood
supply to the heart muscle.

2. Dilatation of the bronchi allowing a greater amount of air to
enter the lungs at each inspiration.

3. Dilatation of the blood vessels to the skeletal muscles increasing
the supply of oxygen and nutritional material to the muscles. This
enables muscle activity to be sustained.

4. Constriction of the blood vessels to the skin, thus raising the blood pressure.

5. Contraction of the spleen, thus increasing the volume of circulating blood.

6. Increasing the rate of change of glycogen to glucose, thus ensuring sufficient glucose for sustained muscle contraction.

7. Dilatation of the pupil of the eye due to stimulation of the radiating muscle fibres of the iris.

8. Slowing down of peristalsis in the digestive tract and limiting the flow of saliva.

9. Increasing the tone of the anal and urethral sphincter muscles, thus inhibiting micturition and defaecation.

10. Increasing the activity of the sweat glands and contraction of the arrectores pilorum causing 'goose flesh'.

Adrenalin and noradrenalin prepare the body to deal with abnormal conditions so that it responds to fear, excitement and danger effectively.

Figure 342

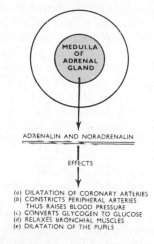

Scheme showing some of the functions of adrenalin and noradrenalin.

The Islets of Langerhans

The cells which make up the Islets of Langerhans are found in clusters irregularly distributed throughout the substance of the pancreas. Unlike the pancreatic tissue which produces a digestive juice there are no ducts leading from the clusters of Islet cells. Their secretion passes directly into the pancreatic veins and so circulates throughout the body in the blood.

There are two main types of cell in the Islets of Langerhans; the α cells which produce a hormone called *glucagon* and the β cells which produce *insulin*. Both hormones influence the level of glucose in the blood.

The normal blood glucose level is between 80 and 120 mg per cent. When there has been excessive exercise or an insufficient intake of carbohydrate foods it may fall below normal. When this happens glucagon has the effect of raising the blood glucose level to normal by mobilising the glycogen stores in the liver.

Insulin has the opposite effect, *it reduces the blood glucose level*. The mode of action of insulin is not fully understood but it has been suggested that it facilitates the passage of the glucose molecule across the cell membrane. Once the glucose molecules are inside the cell they can be katabolised to provide energy for other activities within the cell. In muscle and liver cells, if there is an excess of glucose which is not immediately required to provide energy, it is changed into glycogen and stored there for future use. Insulin also has some, as yet not fully explained, action in relation to the storage and mobilisation of fats in the body.

An insufficiency of insulin in the body leads to the development of *diabetes mellitus* which is characterised by disturbances in both glucose and fat metabolism. Patients suffering from this condition have a blood glucose level which exceeds the normal renal threshold level and so glucose is found in the urine. Because of the high concentration of glucose in the urine an excessive amount of water is excreted leading to dehydration and polydypsia. Excessive amounts of fat are partially metabolised to the stage of keto acids. These accumulate in excess upsetting the acid-base balance of the blood.

Patients suffering from diabetes mellitus are rarely, if ever, cured but they can be effectively maintained by replacement therapy. Insulin is not stored within the body so it has to be given at least daily.

The Thymus Gland

The thymus gland lies in the thoracic cavity immediately behind the sternum and in front of the arch of the aorta, brachiocephalic vein and trachea. Its lower surface is in close association with the pericardium and may reach the level of the fourth costal cartilage. The gland varies in size depending on the age of the individual. During childhood it grows to a fairly large size, at puberty it begins to diminish until in the adult it is quite small and fibrosed.

In early childhood it is pinkish grey in colour and is composed of two lobes connected by areolar tissue. Each lobe is surrounded by a fibrous capsule which penetrates into the substance of the gland partially dividing it into lobules.

The gland is composed of a *cortex* and *medulla*.

CORTEX

This is composed mainly of lymphocytes supported by connective tissue.

MEDULLA

The medulla is composed of some flattened epithelial cells and a few lymphocytes supported by connective tissue.

As the individual grows older the cells are replaced by fatty tissue.

The arterial blood supply to the gland is through the inferior thyroid arteries.

The venous return is into the brachiocephalic vein.

Figure 343

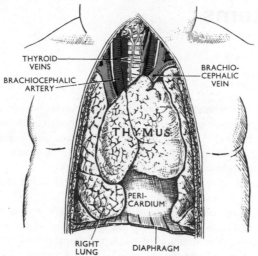

Diagram illustrating the position of the thymus gland in a young child.

FUNCTIONS OF THE THYMUS GLAND

A substance called *thymosin* has been isolated from the thymus which is believed to stimulate immunological activity in lymphoid tissue, especially before puberty.

Before puberty it is a source of lymphocytes.

The Pineal Gland or Body

The pineal gland is a small body situated in the brain below the corpus collosum and posterior to the third ventricle. It is reddish grey in colour, and is approximately 10 mm in length. It is surrounded by a fine capsule and is composed of epithelial cells arranged to form lobules which are surrounded by fine connective tissue.

Its functions are quite unknown.

17. The Reproductive Systems

The ability to reproduce is one of the properties which distinguish living from non-living matter. The more primitive the animal the simpler is the process of reproduction. In human beings reproduction is a very complicated process involving the existence of two sexes, male and female, known as sexual reproduction.

The reproductive organs of the male and female differ anatomically to ensure their satisfactory functional activity.

The initial cell which will eventually form a new life was termed the *zygote*. The zygote is formed by the fertilisation of the ovum by a spermatozoon. The function of the female reproductive system is to form the ovum and after fertilisation to nurture it until it is capable of an independent existence. The function of the male reproductive system is to form and transmit the spermatozoon to the female.

The Female Reproductive System

The female reproductive system or generative organs are divided into two groups:

the external organs of reproduction—the genitalia
the internal organs of reproduction.

THE EXTERNAL ORGANS OR GENITALIA

The external genitalia are known collectively as the *vulva* which consists of several structures:

the labia majora (labum majus)   the vestibule
the labia minora (labum minus)   the hymen
the clitoris   the greater vestibular glands

The labia majora

The labia majora are two large folds containing sebaceous and sweat glands embedded in fibrous tissue and covered with skin. Anteriorly the two folds are joined to form the *mons pubis*.

The mons pubis is an elevation produced by an extra distribution of fatty tissue over the symphysis pubis. The labia majora form the lateral boundaries of the vulva and join posteriorly to form the *posterior commissure*. At puberty the mons pubis and the lateral aspect of the labia become covered with hair.

The labia minora

The labia minora lie within the labia majora. They are two folds of connective tissue covered with skin. Anteriorly they are divided into two parts, one stretching over the clitoris to form the *prepuce*, the other passing beneath it to form the *frenulum*.

Posteriorly the labia minora are fused together forming the *fourchette*, a structure which may be torn at the delivery of the first baby.

Figure 344

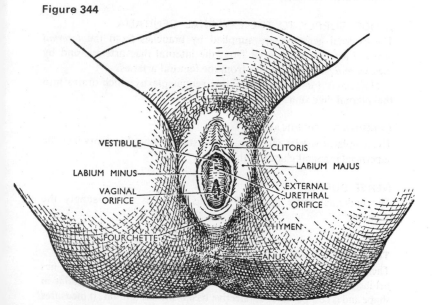

Diagram showing the female external genitalia

The clitoris

The clitoris corresponds to the penis in the male and contains erectile tissue. It is attached to the symphysis pubis by a suspensory ligament and lies between a double fold of the labia minora forming the prepuce above and the frenulum below. It consists of two cylindrical bodies surmounted by a *glans* covered with mucous membrane containing specialised nerve endings.

The vestibule

The vestibule is a smooth almond-shaped area lying between the labia minora into which the vagina, urethra and ducts of the greater vestibular glands open.

The hymen

The hymen is a thin membrane of connective tissue covered by mucous membrane. It partially occludes the opening of the vagina and is usually perforated centrally.

The greater vestibular glands

The greater vestibular glands lie in the labia majora, one on either side at the commencement of the vagina. They are about the size of a small pea and are covered with dense fibrous tissue. There are ducts which lead from the glands and open into the vestibule. The glands secrete mucus which lubricates the vulva.

BLOOD SUPPLY TO THE EXTERNAL GENITALIA

The external genitalia are supplied by branches from the *internal pudendal arteries* which arise from the internal iliac arteries, and by the *external pudendal arteries* from the femoral arteries.

The venous drainage forms a large venous plexus which drains into the internal iliac vein.

LYMPHATIC DRAINAGE

The lymphatic vessels which drain the external genitalia empty into the superficial inguinal glands in the groin.

NERVE SUPPLY

The pudendal nerve gives off various branches which supply the genitalia.

THE PERINEUM

The perineum is the area extending from the fourchette, which lies behind the labia minora, to the anal canal. It is roughly triangular in shape and is composed of connective tissue, muscle and fat. It measures approximately 4 cm by 4 cm ($1\frac{1}{2}$ inches by $1\frac{1}{2}$ inches). It forms an attachment for the muscles of the pelvic floor and occupies the area between the rectum and the vagina.

THE INTERNAL ORGANS OF THE FEMALE REPRODUCTIVE SYSTEM

The internal organs of the female reproductive system lie in the pelvic cavity and consist of:

   the vagina
   the uterus
   the 2 uterine tubes
   the 2 ovaries.

Figure 345

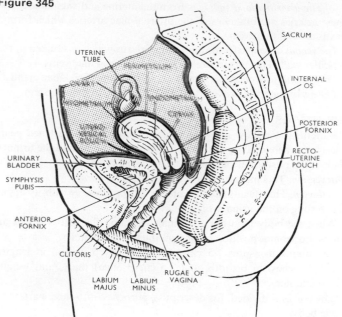

Illustration of the pelvic peritoneum and organs in association with the uterus. Lateral view.

THE VAGINA

The vagina is a fibro-muscular tube connecting the internal and external organs of generation. It runs obliquely upwards and backwards at an angle of approximately 45°. The anterior and posterior walls are in apposition except at the vault where it is divided into *four fornices* by the protruding cervix of the uterus.

The anterior fornix lies in front of the cervix and is in contact with the base of the bladder.

The posterior fornix lies behind the cervix.

The lateral fornices are the areas of the vagina on either side of the cervix where it protrudes into the vagina.

The anterior wall of the vagina measures 6·25 cm (2½ inches) and the posterior wall measures 7·5 cm (approximately 3 inches).

Structure of the vagina

The vagina has three layers of tissue.

1. An outer covering of *areolar and elastic tissue containing* bundles of nerves and many blood vessels.

2. A middle layer of *smooth muscle tissue*.

3. An inner lining of *stratified squamous epithelium* which is arranged in transverse folds, or *rugae*.

The arterial blood supply is from the uterine and vaginal arteries. These arteries are branches of the internal iliac arteries which form a plexus around the vagina.

The venous drainage forms the venous plexus which is situated in the muscular wall. This plexus drains into the internal iliac veins.

The lymphatic drainage is into the deep and superficial iliac glands.

The nerve supply is provided by the autonomic nervous system.

THE UTERUS

The uterus is a hollow muscular organ, shaped like a flattened pear. It lies in the pelvic cavity in front of the rectum and behind the urinary bladder. Its position is one of *anteversion and anteflexion*.

Anteversion. This means that the uterus *leans forwards*.

Anteflexion. This means that the uterus is *bent forwards* with its upper part resting on the urinary bladder.

When the body is in the anatomical position the uterus lies in an almost horizontal position resting on the urinary bladder.

The uterus measures 7·5 cm (approximately 3 inches) in length, 5 cm (2 inches) in breadth, and 2·5 cm (1 inch) thick, and weighs 40–50 grammes.

The uterus is divided, for descriptive purposes, into three parts:
the body
the cervix
the fundus.

The body

The rounded upper part of the body is called the *fundus*. Just below the fundus the uterine tubes open out one on either side. The body then tapers down towards the cervix where there is a constricted portion named the *isthmus*. The cavity of the body is triangular in shape. The uppermost part is the *base* and the lowest part of the triangle is the *apex*. At the apex the anterior and posterior walls are in apposition.

The cervix

The cervix is cylindrical in shape and measures 2·5 cm (1 inch) in length. The cavity of the cervix is spindle shaped, communicating with the isthmus of the body at a constricted part known as the *internal os*, and opening into the vagina at the *external os*.

Part of the cervix protrudes through the anterior vaginal wall with the external os pointing towards the posterior vaginal wall. The part above the vagina is called the *supra-vaginal cervix*, the part protruding into the vagina is the *intra-vaginal cervix*.

Structure of the body of the uterus

The body of the uterus is composed of three layers of tissue:
the perimetrium
the myometrium
the endometrium.

Figure 346

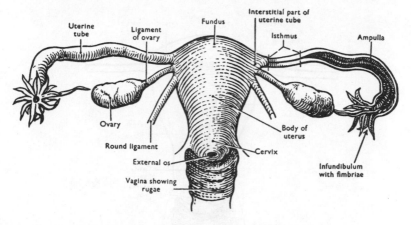

The uterus showing position of uterine tubes and ovaries. Anterior view.

The Perimetrium. This consists of peritoneum. It covers the fundus and the anterior surface to the level of the internal os, and is then reflected on to the bladder forming a small pouch between the uterus and the bladder, this pouch is termed the *utero-vesical pouch*. The posterior surface is covered to where the cervix protrudes into the vagina and is then reflected on to the rectum forming the *recto-uterine pouch*. Laterally the perimetrium extends over the uterine tubes forming a double fold, *the broad ligament,* leaving the lateral borders of the body uncovered.

The Myometrium. This is a thick muscle layer composed of bundles of smooth muscle fibres arranged in three interlacing layers:
    the *inner layer* of fibres runs in a circular fashion
    the *middle layer* of fibres runs obliquely
    the *outer layer* of fibres runs in a longitudinal fashion.

The Endometrium. This is the lining of the uterus. It is composed of columnar epithelium (before puberty it is ciliated). It contains many straight tubular glands.

Structure of the cervix
The cervix is composed of fibrous connective tissue and smooth muscle. It is lined with columnar epithelium which is arranged in folds.

Supports of the uterus
The uterus is supported in the pelvic cavity by a number of structures:
    the surrounding organs
    the muscles of the pelvic floor
    four pairs of ligaments derived from folds of peritoneum and
        connective tissue.

Figure 347

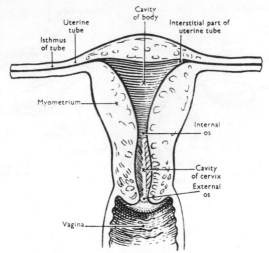

The interior of the uterus.

1. *Two Broad Ligaments*. These are formed by double folds of peritoneum on either side of the uterus. They hang downwards from the uterine tubes and are attached to the sides of the pelvis.

2. *Two Round Ligaments*. These are bands of fibrous tissue enclosed in layers of peritoneum and are attached to the body of the uterus below the uterine tubes. They pass between the folds of the broad ligament to the sides of the pelvis and then forwards through the inguinal canal and fuse with the labia majora.

Figure 348

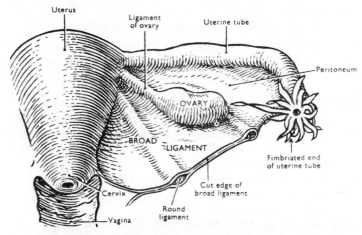

Scheme showing the formation of the broad ligament.

3. *Two Transverse Cervical Ligaments or Cardinal Ligaments.* These ligaments are attached to the sides of the cervix and extend laterally to the walls of the pelvis.

4. *Two Utero-sacral Ligaments.* These ligaments are attached to the upper part of the posterior wall of the cervix and extend backwards on either side of the rectum to be inserted into the sacrum.

Arterial blood supply to the uterus
The main blood supply to the uterus is from the *uterine arteries* which are branches of the internal iliac arteries. They run up the lateral borders of the uterus and anastomose with the ovarian arteries just below the level of the uterine tubes. The blood vessels are not straight but run a tortuous course twisting in and out through the muscle fibres.

Venous drainage
There is a plexus of veins which lies between the layers of the broad ligament and which drains blood from the uterus into the uterine and ovarian veins.

Lymphatic drainage
The main lymphatic drainage is through the deep and superficial iliac glands.

Nerve supply
The nerve supply is through the sympathetic and parasympathetic nerves. The sympathetic nerves are derived from the hypogastric plexus, and the parasympathetic from the first, second and third sacral nerves.

Functions of the uterus
The functions of the uterus are:
1. to receive the fertilised ovum, to nourish and protect it during its development and to expel it at full term
2. to be involved in the menstrual cycle.

THE TWO UTERINE TUBES
The uterine tubes lie one on each side of the uterus in the folds of the broad ligament. Opening from the uterus they pass outwards and backwards to penetrate the posterior wall of the broad ligament, and open into the peritoneal cavity in close relationship to the ovaries. They measure about 10 cm (4 inches) in length, and are described in four parts.

1. *The Interstitial Portion.* This part lies within the wall of the uterus between the fundus and the body.

2. *The Isthmus.* This is a straight narrow part just lateral to the wall of the uterus.

3. *The Ampulla.* This is the widest part of the tube and is long and tortuous.

4. *The Infundibulum.* This is a dilated trumpet-like portion which opens into the peritoneal cavity. The end of the tube has finger-like projections called *fimbriae,* one of which is longer than the others and is called the *ovarian fimbria.*

Figure 349

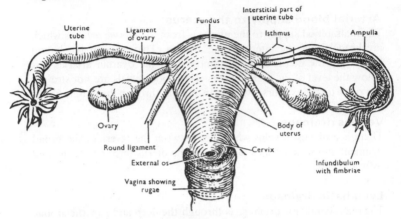

Diagram showing the parts of a uterine tube.

Structure of the uterine tubes
The uterine tubes are thin muscular tubes composed of three layers of tissue.

1. *An inner lining of ciliated epithelium,* thrown into longitudinal folds which almost obliterate the lumen of the tube.

2. *A middle coat composed of muscular tissue* which is a continuation of the uterine muscle. The muscle coat in the uterine tubes is much thinner than that of the uterus, and has only two layers of fibres, a longitudinal layer and a circular layer.

3. *An outer layer which consists of peritoneum* is formed by the broad ligament. The peritoneum covers the superior, anterior and posterior surfaces of the tubes, but not the inferior surface.

The blood supply
This is from the *uterine* and the *ovarian arteries.*

Function
The function of the uterine tubes is to convey the ovum from the ovary in the peritoneal cavity to the uterus.

THE OVARIES
The ovaries are the female gonads or sex glands. They lie in a shallow fossa on the lateral walls of the pelvis, and are attached to the posterior

layer of the broad ligament by a band of peritoneum known as the *mesovarium*.

The size of the ovaries varies in different individuals. Their length varying from 2·5 cm to 3·5 cm, their breadth is about 2 cm and they are 1·25 cm thick.

Structure of the ovaries

The ovaries are divided into a cortical zone and a medulla.

The medulla is composed of fibrous tissue and contains blood vessels, lymphatic vessels and nerves. These structures enter and leave the ovaries at the *hilum*.

The cortex is composed of a framework of connective tissue, called the *stroma*, in which lie numerous follicles, and it is covered with cubical cells of epithelium called *the germinal epithelium*.

Figure 350

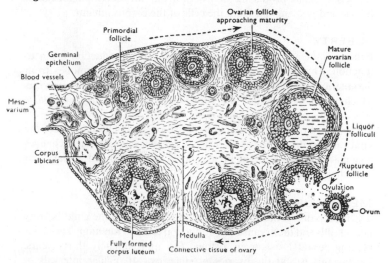

Interior of an ovary showing the development of an ovarian follicle. The arrows indicate the order of the stages of development.

The appearance of the cortex depends on the age of the individual and the state of the menstrual cycle. Before puberty the surface is smooth and the follicles in the stroma are small and immature and are termed *primordial follicles*. During the child-bearing period at each menstrual cycle one or more of these follicles are stimulated to grow. Each of these *ovarian follicles* contains a maturing ovum. At each cycle usually one follicle matures and rises to the surface of the ovary, ruptures and releases the ovum into the peritoneal cavity. The remains of the follicle is further stimulated and develops as a *corpus luteum* (yellow body). If the ovum is not fertilised the corpus luteum degenerates and is transformed gradually into a scar-like white body called the *corpus albicans* (white body) by the end of the cycle. After the menopause there is a gradual disappearance of the follicles but white bodies abound throughout the stroma.

Arterial blood supply
This is from the *ovarian arteries* which are branches of the abdominal aorta.

Venous drainage
This is into the venous plexus which lies behind the uterus. The ovarian veins arise from this plexus. The right ovarian vein opens directly into the inferior vena cava and the left, into the left renal vein.

Functions
The function of the ovaries is to produce the ova (egg cells). It is also an endocrine gland producing *oestrogen* from the developing ovarian follicle and *progesterone* from the cells of the corpus luteum.

PUBERTY
Puberty is the age when the internal organs of reproduction begin to function. The ovaries are stimulated at this time by hormones from the anterior lobe of the *pituitary gland*. There are two of these hormones which together are called *the gonadotrophins*. They are the *follicle stimulating and luteinising hormones*.

The age of puberty varies in different races and in individuals of the same race. In Britain the average age is thirteen years. Physical changes take place at this time. The breasts begin to develop and axillary and pubic hair begins to grow.

MENOPAUSE
The menopause is the term applied to the time when the child-bearing period ends and the organs of reproduction cease to function. The child-bearing period lasts on an average for approximately thirty years. During this period the ovaries produce ova at regular intervals of about 28 days and changes take place in the uterus in preparation for a possible pregnancy. This pattern of events is known as the *menstrual cycle*.

THE MENSTRUAL CYCLE
The menstrual cycle involves a series of events which occur at regular intervals over a period of about 28 days involving three distinct phases:
>   *the proliferative phase* . . . *14 days.*
>   *the secretory phase* . . . *10 days.*
>   *the menstrual phase* . . . *4 days.*

The proliferative phase
During this phase the anterior lobe of the *pituitary gland* secretes the

Figure 351

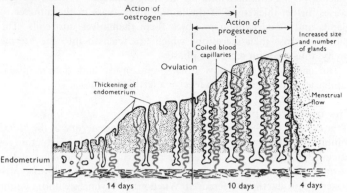

Scheme of the changes which occur in the uterus during the menstrual cycle.

follicle stimulating hormone. This hormone circulating in the blood stream has a local effect upon the ovary stimulating the growth and development of an *ovarian follicle.* Within the follicle lies an ovum which also grows and develops. During this period usually only one ovum grows, develops and matures, but sometimes two or more may do so. When the ovum is mature the follicle ruptures and the ovum is cast out into the peritoneal cavity. The release of the ovum from the follicle is termed *ovulation.* The ovum is then caught up by the fimbriated end of the uterine tube and propelled along the tube to the uterus, possibly by peristalsis and the movement of the cilia of the lining membrane of the uterine tube.

During the growth and development of the ovarian follicle the cells secrete a hormone *oestrogen.* This hormone passes directly into the blood stream and has its effect upon the endometrium; stimulating the columnar cells to reproduce more rapidly, forming a thicker lining and increasing the number of glands and blood capillaries.

The secretory phase

Immediately after ovulation the walls of the ovarian follicle collapse. Due to the high level of oestrogen in the blood at this time, the follicle stimulating hormone from the pituitary gland is withdrawn and the pituitary gland starts to secrete the second gonadotrophin, the *luteinising hormone.*

This hormone now circulating in the blood stimulates the remaining cells of the ovarian follicle to grow rapidly and become filled with a yellow substance known as *lutein.* This new structure is known as the *corpus luteum* or yellow body. The cells of the corpus luteum under the influence of the luteinising hormone start to secrete the second ovarian hormone *progesterone.*

Progesterone is directly absorbed into the blood stream and has an effect upon the endometrium, augmenting the action of oestrogen. It causes further thickening of the lining, an increase in the number of glands and blood capillaries which form themselves into coils. The glands increase their secretions which are released into the uterine cavity, thus the term secretory phase.

The endometrium is now thick, soft, moist and rich in blood capillaries and is prepared to receive a fertilised ovum.

The menstrual phase

If the ovum is not fertilised, the high level of progesterone in the blood inhibits the activity of the pituitary gland and the production of luteinising hormone is considerably reduced. The withdrawal of this hormone causes degeneration of the corpus luteum and thus progesterone production is decreased. When progesterone is withdrawn the endometrium within the uterus breaks down, the cells lining the uterus

Figure 352

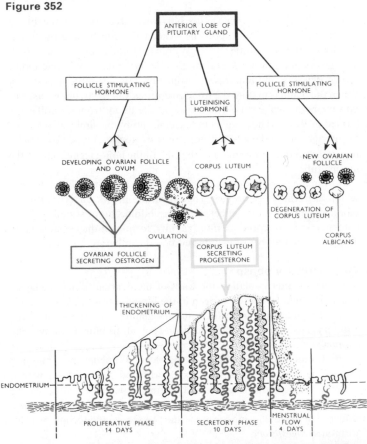

Schematic drawing showing the hormonal control of the menstrual cycle.

disintegrate and the capillaries break down. The extra secretions, plus lining cells, blood from the broken down capillaries and the unfertilised ovum constitute the *menstrual flow.*

When the progesterone level in the blood falls to a certain level the pituitary gland again starts to secrete the follicle stimulating hormone, another ovarian follicle is stimulated and the cycle starts again.

Note. If the ovum is fertilised there is no breakdown of the endometrium and no menstrual flow. The fertilised ovum embeds itself within the endometrium and produces a hormone named *chorion gonadotrophin* which keeps the corpus luteum intact, and thus allows it to continue secreting progesterone.

The fertilised ovum during the first 12 weeks is called the *embryo,* and thereafter the *fetus.* It is protected and nourished for 40 weeks (nine months) within the uterus and is then said to be at *full term,* and sufficiently mature to lead an independent existence and is ready to be born.

THE BREASTS OR MAMMARY GLANDS

The breasts or mammary glands are accessory glands of the female reproductive system. They exist in the male as well as in the female, but only in a rudimentary form.

They are described as large hemispherical eminences extending from the *second to the sixth rib,* and from the sides of the sternum to the midline of the axillae. Before puberty they are very small but develop

Figure 353

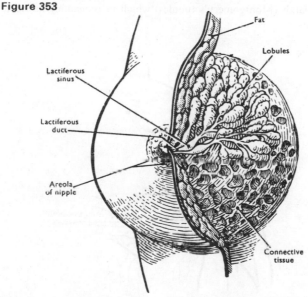

Lactiferous sinus

Lactiferous duct

Areola of nipple

Fat

Lobules

Connective tissue

Interior of the breast.

after puberty and are still further increased in size during pregnancy and lactation. In old age they atrophy.

Structure of the mammary glands

The mammary glands consist of the following tissues:

glandular tissue
fibrous tissue
fatty or adipose tissue.

The Glandular Tissue. This consists of about twenty lobes in each breast. Each lobe is made up of lobules. The lobules consist of a cluster of alveoli which open into small ducts which unite to form large excretory ducts called the *lactiferous ducts.* The lactiferous ducts converge towards the centre of the breast where they form dilatations or reservoirs for milk during lactation. From these dilatations run very narrow ducts which open out on to the surface at the nipple.

The Fibrous Tissue. This encloses the glandular tissue and forms the supporting suspensory ligaments.

The Fatty or Adipose Tissue. This covers the surface of the gland and is found between the lobes; the amount of fatty tissue determines the size of the breasts.

At the centre of the outer surface is a small conical eminence, *the nipple.* The base of the nipple is surrounded by a pigmented area, *the areola.* This varies in colour from a deep pink to a light brown colour. On the surface of the areola there are numerous sebaceous glands called areolar glands (Montgomery's tubules) which in pregnancy lubricate the nipple.

Figure 354

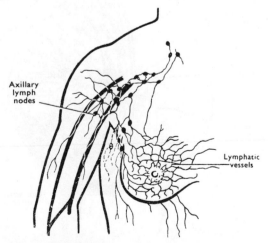

Axillary lymph nodes

Lymphatic vessels

Lymphatic drainage of the breast.

Arterial Blood Supply. The breasts are supplied with blood from the thoracic branches of the axillary arteries and from the internal mammary and intercostal arteries.

Venous Drainage. This describes an anastomotic circle round the base of the nipple called the circulus venosus. Branches from this carry the venous blood to the circumference and end in the axillary and mammary veins.

Lymphatic Drainage. This is mainly into the axillary lymphatic glands. glands.

Functions of the mammary glands

The mammary glands are active only during pregnancy and after the birth of a baby.

In pregnancy they produce a fluid called *colostrum,* which is present until two or three days following the birth of the baby. They are then stimulated by a hormone known as *prolactin* which is secreted from the anterior lobe of the *pituitary gland.* This hormone stimulates the breasts to secrete milk.

The Male Reproductive System

The male reproductive system consists of the following organs:

two testes   ⎱ within the scrotum
two epididymides ⎰
two deferent ducts and spermatic cords
two seminal vesicles
two ejaculatory ducts
the prostate gland
the penis.

THE SCROTUM

The scrotum is a pouch containing the testes, the epididymides and lower part of the spermatic cords. It lies below the symphysis pubis and in front of the upper parts of the thighs and behind the penis.

It is covered with skin which is deeply pigmented and thrown into folds. Beneath the skin lies a thin sheet of involuntary muscle—the *dartos muscle.* This muscle forms a septum dividing the scrotum into two cavities, one testis lying in each cavity. Interior to the dartos muscle lies the *cremaster muscle* and its fascia.

THE TESTES

The testes are the reproductive glands of the male and are the equivalent of the ovaries in the female. They are suspended in the scrotum by the spermatic cord. The testes are approximately 4·5 cm in length, 2·5 cm in breadth and 3 cm in thickness.

The testes are surrounded by three distinct layers of tissue.

Figure 355

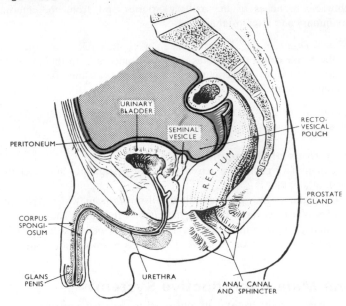

Illustration of the pelvic peritoneum and pelvic organs in the male.

The tunica vaginalis

This is the outer covering of the testes and lines the cremaster muscle. It is in a double fold forming a *parietal and visceral* layer.

The tunica vaginalis is a downgrowth of the abdominal and pelvic peritoneum. During early fetal life the testes develop in the lumbar

Figure 356

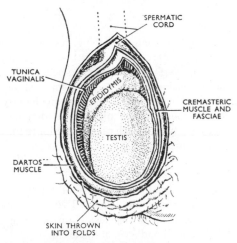

Diagram showing the structure of the scrotum.

region then descend into the scrotum pushing a piece of peritoneum with them. This peritoneum eventually surrounds the testes in the scrotum and is normally cut off from the remainder of the abdominal and pelvic peritoneum.

The tunica albuginea
This is a fibrous covering surrounding the testes and situated under the visceral layer of vaginalis. Ingrowths of the tunica albuginea form septa dividing the glandular structure of the testes into *lobules*.

The tunica vasculosa
This consists of a network of capillaries supported by delicate connective tissue which lines the tunica albuginea. Therefore each lobule is surrounded by a fine network of blood capillaries.

Figure 357

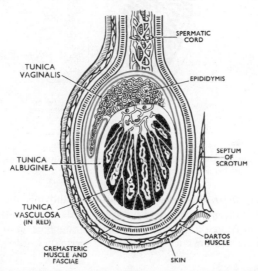

Diagram illustrating the structure of the testis, and its coverings.

Structure of the testes
Each testis consists of from two to three hundred *lobules* composed of germinal epithelial cells which are formed into the *convoluted seminiferous tubules*. Between the tubules are found groups of secretory cells known as the *interstitial cells*. The tubules eventually straighten out to become the straight seminiferous tubules. The straight tubules ascend to the upper pole of the testis, many of them joining together to become the *efferent ducts* which again join to become a complicated tortuous tubule, *the epididymis*.

The epididymis is a convoluted tubule which is folded upon itself and descends to the lower pole of the testis lying posterior to the seminiferous tubules. It is tightly packed and supported by fine areolar tissue. At the lower border of the testis the epididymis continues as the *ductus deferens*.

Figure 358

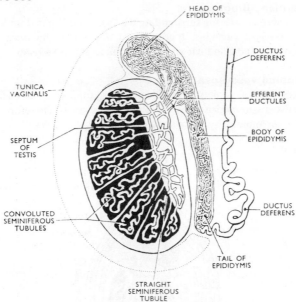

Diagram showing the arrangement of the seminiferous tubules, epididymis and ductus deferens.

The ductus deferens is a continuation of the epididymis. It commences at the tail of the epididymis and passes upwards on the posterior wall of the testis. At first it is a coiled tube then straightens out to leave the testis and the scrotum enclosed within the spermatic cord.

Figure 359

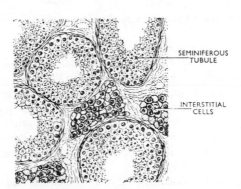

Microscopic structure of the testis.

THE SPERMATIC CORDS

There are *two spermatic cords*, one from each testis. Each spermatic cord is composed of the following structures:

the testicular artery

the testicular vein

lymphatic vessels

nerves

the ductus deferens.

The spermatic cord suspends the testis in the scrotum. It is composed of a thin sheet of fibrous tissue which covers an inner layer of muscle tissue. Fine connective tissue surrounds the blood vessels, nerves and ductus deferens.

The spermatic cord passes through the inguinal canal. At the deep inguinal ring the structures within the cord diverge.

The testicular artery is a branch of the abdominal aorta and arises from it just below the renal arteries.

The testicular vein passes upwards through the pelvic cavity, the left vein opening into the left renal vein, and the right vein opening into the inferior vena cava.

The lymphatic drainage from the testis is into the pre-aortic lymph nodes.

The ductus deferens passes upwards from the testis through the inguinal canal and ascends medially towards the posterior wall of the bladder where it is joined by the duct from the *seminal vesicle* to form the *ejaculatory duct*.

THE SEMINAL VESICLES

The seminal vesicles are two pouches which lie on the posterior aspect of the bladder. They are approximately 5 cm in length and roughly pyramidal in shape.

The seminal vesicles are composed of:

an outer coat of fibrous tissue

a middle coat of muscular tissue

an inner lining of columnar epithelial tissue.

The lower part of the seminal vesicles narrow and open into two short ducts which join with the deferent ducts to form the ejaculatory ducts.

THE EJACULATORY DUCTS

The ejaculatory ducts are two short tubes approximately 2 cm in length. They pass forward and downward to open into the *prostate gland*.

The ejaculatory ducts are composed of the same layers of tissue as the seminal vesicles, i.e., fibrous, muscular and columnar epithelial tissue.

THE PROSTATE GLAND

The prostate gland lies in the pelvic cavity immediately in front of the

Figure 360

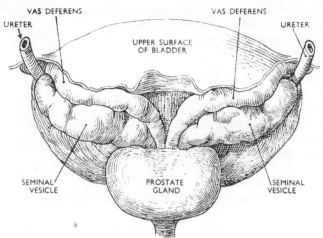

View of the organs in association with the bladder from behind.

rectum and just behind the symphysis pubis. It *surrounds* the commencement of the urethra. The gland is about the size of a chestnut. It is described as having a *base* above and an *apex* below.

Structure

An outer capsule of *fibrous tissue*.

Immediately beneath the capsule there is involuntary muscle tissue which penetrates deeply into the glandular substance:

The glandular substance is composed of *columnar epithelium* arranged in *follicles*. The epithelial cells line the many minute ducts which open into the urethra.

Note. Enlargement of the prostate gland commonly occurs in men over middle age. This enlargement produces pressure on the urethra, thus stopping the flow of urine resulting in *retention of urine.*

THE URETHRA AND PENIS

Urethra

The male urethra forms a common pathway for the flow of urine and the flow of secretions from the male reproductive organs (the semen). It is much longer than the female urethra and measures 19 to 20 cm (7·5 to 8 inches). It extends from *the internal urethral sphincter* of the urinary bladder to the extremity of the penis.

It passes through the prostate gland, this part of the urethra is known as the *prostatic urethra*. Immediately below this is situated the *external urethral sphincter*. The urethra then curves at an angle of 90° through the perineum then passes downwards and is surrounded by the structures of the penis.

Figure 361

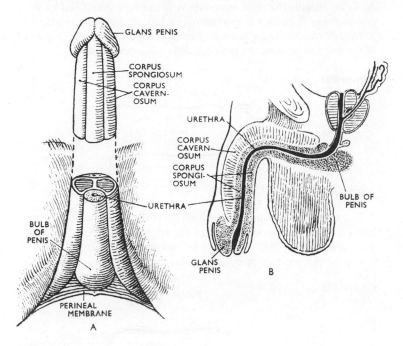

The penis.

A. Under surface   B. Side view.

Penis

The penis is composed of a *root* and a *body*. The root lies in the perineum and the body surrounds the urethra.

The penis is formed by three elongated masses of *erectile tissue* and involuntary muscle very rich in blood vessels. The erectile tissue is supported by fibrous tissue and covered with skin which is deeply pigmented.

The two lateral columns of tissue are known as the *corpora cavernosa*.

The middle column is termed the *corpus spongiosum* and lies between the two lateral columns.

At the extremity or tip of the penis the corpus cavernosum is expanded into a triangular structure known as the *glans penis*. Just above the glans the skin is folded upon itself and forms a movable fold known as the *foreskin* or *prepuce*.

Note. The foreskin in some male infants is too tight, resulting in difficulty in micturition. This condition is known as a *phimosis* and to relieve it the surgeon performs a circumcision.

FUNCTIONS OF THE MALE REPRODUCTIVE SYSTEM

As in the female, the male reproductive organs are stimulated by the *gonadotrophic hormones* from the anterior lobe of the pituitary gland.

The follicle stimulating hormone stimulates the *seminiferous tubules* of the testes to produce the male egg cells—the spermatozoa.

Figure 362

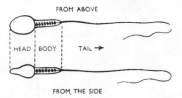

A spermatozoon.

The spermatozoa consist of a head, a body and a tail. They are conveyed along the tubules, through the epididymides to the deferent duct. Within the epididymides they become motile.

The spermatozoa are then conveyed to the seminal vesicles where a

Figure 363

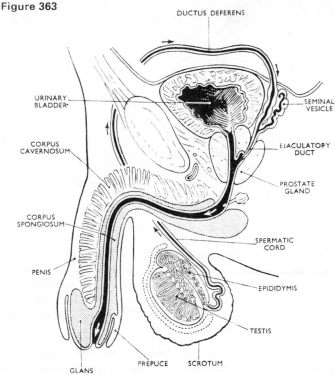

Scheme showing the flow of semen and spermatozoa.

thin viscid secretion is added. The spermatozoa plus this viscid fluid form the *semen*.

The semen flows through the ejaculatory ducts to the prostate gland. The secretion from the prostate gland is thought to act as a lubricant for the passage of the semen through the urethra.

The interstitial cell stimulating hormone (luteinising hormone) stimulates *the interstitial cells* of the testes to produce the hormone *testosterone*.

Testosterone is absorbed directly into the blood stream. Under its influence the male reproductive organs develop and function satisfactorily.

Testosterone is also responsible for the changes in the male which occur at puberty, that is, the breaking of the voice, development of the reproductive organs, growth of hair on the face, chest, axillae and pubis. It is also responsible for the satisfactory functioning of the seminiferous tubules ensuring their ability to produce healthy spermatozoa.

When the spermatozoa are introduced into the female reproductive system they fertilise an ovum forming a zygote, which grows, and by the process of mitotic reproduction develops into a new human being.

Bibliography

Adrian, E. D. (1949) *The Basis of Sensation,* reprint. London: Hofner.

Bell, G. H., Davidson, J. N. and Emslie-Smith, D. (1972) *Textbook of Physiology and Biochemistry,* 8th ed. Edinburgh and London: Churchill Livingstone.

Best, C. H. and Taylor, N. B. (1958) *The Living Body,* 4th ed. reprint. London: Chapman and Hall.

Best, C. H. and Taylor, N. B. (1967) *The Physiological Basis of Medical Practice,* 8th ed. Edinburgh and London: Churchill Livingstone.

Cantarow, A. and Schepartz, B. (1967) *Textbook of Biochemistry,* 4th ed. London: Saunders.

Davidson, Sir Stanley, Passmore, R. and Brock, J. F. (1972) *Human Nutrition and Dietetics,* 5th ed. Edinburgh and London: Churchill Livingstone.

Davson, H. and Eggleton, Grace, eds. (1968) *Principles of Human Physiology,* 14th ed. Edinburgh and London: Churchill Livingstone.

Ganong, W. F. (1969) *Review of Medical Physiology.* Los Altos, California: Lange.

Ham, A. W. and Leeson, T. S. (1969) *Histology,* 6th ed. London: Pitman.

Jamieson, E. B. (1950) *A Companion to Manuals of Practical Anatomy,* 7th ed. London: Oxford University Press.

Kimber, D. C. and Gray, C. E., Rev. by Stackpole, C. E. and Leavell, L. C. (1966) *Textbook of Anatomy and Physiology,* 15th ed. New York: Macmillan.

Krech, D. and Crutchfield, R. S. (1969) *Elements of Psychology,* 2nd ed. New York: Knopf.

Lippold, O. C. J., Winton, F. R. and Bayliss, L. E. (1972) *Human Physiology,* 8th ed. Edinburgh and London: Churchill Livingstone.

Romanes, G. J., ed. (1972) *Cunningham's Textbook of Anatomy,* 11th ed. London: Oxford University Press.

Warwick, R. and Williams, P. L., eds. (1973) *Gray's Anatomy,* 35th ed. London: Longman.

Weir, D. M. (1971) *Immunology for Undergraduates,* 2nd ed. Edinburgh and London: Churchill Livingstone.

Wilson, A. and Schild, H. O. (1968) *Applied Pharmacology,* 10th ed. reprint. Edinburgh and London: Churchill Livingstone.

Index

73

PT313 (821)